# CEREBRAL DURAL ARTERIOVENOUS FISTULAS

# CEREBRAL DURAL ARTERIOVENOUS FISTULAS

Edited by

**BRADLEY A. GROSS**

Assistant Professor, Department of Neurological Surgery, University of Pittsburgh, Pittsburgh, PA, United States

**FELIPE C. ALBUQUERQUE**

Department of Neurosurgery, Barrow Neurological Institute, St. Joseph's Hospital and Medical Center, Phoenix, AZ, United States

**BRIAN T. JANKOWITZ**

Associate Professor of Neurosurgery and Cooper University in Camden, NJ

**CAMERON G. MCDOUGALL**

Professor of Neurosurgery, Department of Neurosurgery, Johns Hopkins University School of Medicine, Baltimore, MD, United States

ELSEVIER

ACADEMIC PRESS

An imprint of Elsevier

Academic Press is an imprint of Elsevier
125 London Wall, London EC2Y 5AS, United Kingdom
525 B Street, Suite 1650, San Diego, CA 92101, United States
50 Hampshire Street, 5th Floor, Cambridge, MA 02139, United States
The Boulevard, Langford Lane, Kidlington, Oxford OX5 1GB, United Kingdom



**Notices**
Knowledge and best practice in this field are constantly changing. As new research and experience broaden our understanding, changes in research methods, professional practices, or medical treatment may become necessary.

Practitioners and researchers must always rely on their own experience and knowledge in evaluating and using any information, methods, compounds, or experiments described herein. In using such information or methods they should be mindful of their own safety and the safety of others, including parties for whom they have a professional responsibility.

To the fullest extent of the law, neither the Publisher nor the authors, contributors, or editors, assume any liability for any injury and/or damage to persons or property as a matter of products liability, negligence or otherwise, or from any use or operation of any methods, products, instructions, or ideas contained in the material herein.

**Library of Congress Cataloging-in-Publication Data**
A catalog record for this book is available from the Library of Congress

**British Library Cataloguing-in-Publication Data**
A catalogue record for this book is available from the British Library

ISBN 978-0-12-819525-3



*Publisher:* Nikki Levy
*Acquisitions Editor:* Natalie Farra
*Editorial Project Manager:* Sam W. Young
*Production Project Manager:* Maria Bernard
*Cover Designer:* Miles Hitchen

Typeset by SPi Global, India

# Contents

# Contributors

**Felipe C. Albuquerque**
Department of Neurosurgery, Barrow Neurological Institute, St. Joseph's Hospital and Medical Center, Phoenix, AZ, United States

**Jacob F. Baranoski**
Department of Neurosurgery, Barrow Neurological Institute, St. Joseph's Hospital and Medical Center, Phoenix, AZ, United States

**David I. Bass**
Department of Neurological Surgery, University of Washington, Seattle, WA, United States

**Justin M. Caplan**
Department of Neurosurgery, Johns Hopkins University School of Medicine, Baltimore, MD, United States

**Joshua S. Catapano**
Department of Neurosurgery, Barrow Neurological Institute, St. Joseph's Hospital and Medical Center, Phoenix, AZ, United States

**Tyler S. Cole**
Department of Neurosurgery, Barrow Neurological Institute, St. Joseph's Hospital and Medical Center, Phoenix, AZ, United States

**Rose Du**
Department of Neurosurgery, Brigham and Women's Hospital and Harvard Medical School, Boston, MA, United States

**Andrew F. Ducruet**
Department of Neurosurgery, Barrow Neurological Institute, St. Joseph's Hospital and Medical Center, Phoenix, AZ, United States

**Bradley A. Gross**
Department of Neurological Surgery, University of Pittsburgh, Pittsburgh, PA, United States

**Jawad M. Khalifeh**
Department of Neurosurgery, Johns Hopkins University School of Medicine, Baltimore, MD, United States

**Jennifer E. Kim**
Department of Neurosurgery, Johns Hopkins University School of Medicine, Baltimore, MD, United States

**Michael T. Lawton**
Department of Neurosurgery, Barrow Neurological Institute, St. Joseph's Hospital and Medical Center, Phoenix, AZ, United States

**Michael R. Levitt**
Department of Neurological Surgery; Department of Radiology; Department of Mechanical Engineering; Stroke & Applied Neuroscience Center, University of Washington, Seattle, WA, United States

**L. Dade Lunsford**
Department of Neurological Surgery, University of Pittsburgh Medical Center, Pittsburgh, PA, United States

**Cameron G. McDougall**
Department of Neurosurgery, Johns Hopkins University School of Medicine, Baltimore, MD, United States

**Rajeev D. Sen**
Department of Neurological Surgery, University of Washington, Seattle, WA, United States

**Daniel A. Tonetti**
Department of Neurological Surgery, University of Pittsburgh Medical Center, Pittsburgh, PA, United States

**Robert T. Wicks**
Department of Neurosurgery, Johns Hopkins University School of Medicine, Baltimore, MD, United States

**Christopher C. Young**
Department of Neurological Surgery, University of Washington, Seattle, WA, United States

CHAPTER 1

# Dural arteriovenous fistulas: An introduction and historical perspective

**Bradley A. Gross**
Department of Neurological Surgery, University of Pittsburgh, Pittsburgh, PA, United States

As arteriovenous shunts existing within the dura mater leaflets, dural arteriovenous fistulas (dAVFs) are unique to the neuraxis. Often supplied by a myriad of dural/meningeal vessels, their radiographic and angiographic appearance may seem daunting or enigmatic to the unfamiliar. Paradoxically, however, dural arteriovenous fistula anatomy and pathophysiology is quite simple to understand if one begins with the premise that the problem is a "live" piece of dura with shunting within the leaflets (Fig. 1). Tiny dural arteries "crawl" along the outer and/or inner leaflets of the dura to supply this shunt. A thorough discussion of the typical "menu" of dural arterial feeders is presented in the following chapter.

The shunt can either drain into a vein contained entirely within the dura (a venous sinus) or a vein that exits the dura ("cortical" or spinal vein). This sensible anatomic dichotomy also lays the groundwork for the natural history of the shunt, reviewed in detail in a dedicated chapter. Noninvasive angiographic CT or MR imaging may depict, at times, nonspecific preponderance of tiny meningeal vessels in the vicinity of the involved dura. For fistulas with sinus drainage that may present with attributable symptoms such as tinnitus, this preponderance of vessels that may otherwise be nonspecific will alert the keen interpreter to a potential shunt. For fistulas with cortical drainage, a tangle of "out of place" veins without enlarged arterial feeders should alert the interpreter to a potential dural shunt as opposed to a pial arteriovenous malformation (AVM), all reviewed in a dedicated chapter to neuroimaging and diagnosis.

The focus of the remainder of this text is dedicated to their treatment. Successful treatment is predicated on complete obliteration of the venous outflow. Uniquely, treatment can be targeted to the venous outlet without necessarily the need to devascularize all arterial inflow, in contradistinction

*Cerebral Dural Arteriovenous Fistulas*
https://doi.org/10.1016/B978-0-12-819525-3.00009-5

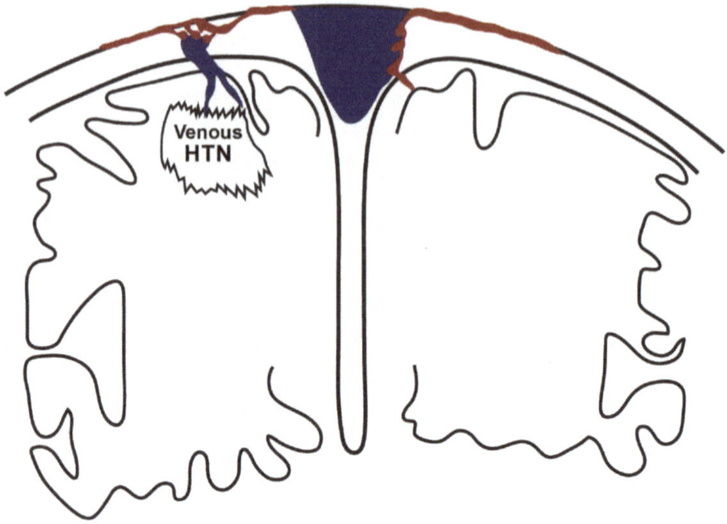


**Fig. 1** Dural arteriovenous fistulas.

to pial AVMs. Exciting advances in endovascular therapy have made it the mainstay of treatment, with basic tenets highlighted in Table 1. Transarterial approaches, typically using liquid embolics, are well suited for nonsinus-based fistulas. Transvenous approaches, that may use occlusive coils and/or liquid embolics, are suitable for any fistula if affording a route to the fistula site. Surgical occlusion remains an excellent but more invasive therapeutic option with high rates of obliteration. At times, hybrid surgical and endovascular treatment may be necessary. Finally, radiosurgery serves as a minimally invasive approach with respectable obliteration rates and serves as an excellent option for appropriately selected lesions. Each of these modalities is presented in detail in respective chapters by highly experienced experts with each approach.

## Historical perspective

The first anatomic description of a dural arteriovenous shunt was provided by Francesco Rizzoli, describing the case of a 9-year-old girl, Giulia, with seizures and occipital pulsatile swelling in 1873.[1] In his monograph, "The diagnosis and treatment of brain tumors," Ernest Sachs describes ligating a "telangiectasia of the dura" in 1931. He is generally credited with the first angiographic description of a dAVF, though angiographic images were lacking.[2]

**Table 1** Basic tenets of dAVF embolization.

| Concept | Pearl/pitfall |
|---|---|
| Transarterial embolization | – The middle meningeal artery is often an excellent first choice conduit<br>– The occipital artery can be deceptively alluring as an arterial conduit but is generally less rewarding than the penetration via MMA. This may change with increasing usage of dual lumen balloon microcatheters<br>– Dual-lumen balloon microcatheter usage, when feasible, can facilitate effective transarterial penetration<br>– In addition to casting the draining vein, *embolisate reflux into other arterial pedicles is an important intraprocedural finding* to suggest effective embolization. Absence of this finding should be concerning for incomplete embolization |
| Transvenous embolization | – Transvenous routes should often be sought and provide direct access to the main target<br>– Mainstay of CCF treatment; be sure to start in the SOV or "past" the fistula to obliterate concerning outflow pathways<br>– Coiling often provides a very safe obliterative option<br>– Liquid embolics may be used, but careful attention to reflux into cavernous ICA, petrosal MMA, or other functional branches should be scrutinized |
| Achieving obliteration | – Significant embolisate should occlude the draining vein. Without this finding, the fistula is still alive and recruitment of additional arterial pedicles should be expected<br>– Full angiography immediately after embolization should be performed to ensure no development of collateral meningeal supply—be particularly weary of contralateral supply<br>– True obliteration is defined as embolisate casting of the draining vein without shunting on full 6-vessel angiography at least 3 months after embolization |

Anatomically stratified case series of dAVFs began to emerge in the 1960s with reference to distinctive indirect/dural carotid-cavernous fistulas[3, 4] and particularly transverse-sigmoid sinus-based fistulas.[5, 6] Houser et al. reviewed the Mayo clinic experience from 1958 to 1971 with 28 cases of dAVFs and were the first to clearly present the significant of cortical venous drainage[7]:

> *Intracranial hemorrhages occurred in those patients in whom the venous drainage of the arteriovenous malformation was limited to pial veins....*
>
> *If the venous outflow was antegrade through the usual channels, the clinical syndrome reflected only the presence and volume of the arteriovenous shunt.*

## It's Djindjian's classification

In his timeless text, *Superselective Angiography of the External Carotid Artery*, Rene Djindjian elegantly depicts relevant meningeal arterial anatomy and provides a modernized discussion of the anatomy and treatment of cerebral dAVFs that is far ahead of its time.[8] The angiographic quality for the era was phenomenal, and his depiction of types of dAVFs has withstood the test of time and remains the mainstay of classification today. The text is a "must read" for any cerebrovascular practitioner.

Djindjian describes four types of dAVFs:

**(1)** Pure meningeal arteriovenous fistulae draining into a sinus or meningeal vein.

**(2)** Pure meningeal arteriovenous fistulae draining directly into a sinus but with reflux into veins discharging into the sinus. It is noted that these are associated with neurological complications unlike the first type.

**(3)** Pure meningeal arteriovenous fistulae draining into cortical veins.

**(4)** Pure meningeal arteriovenous fistulae with large venous lakes.

Djindjian's pupil, Jean-Jacques Merland, subsequently refined this scheme as the senior author on a paper with Christophe Cognard, published in *Radiology* in 1995, that adds a fifth type of fistula with spinal perimedullary drainage.[9] Type 2 fistulae are also stratified by the absence or presence of cortical reflux.

In the same year, in the *Journal of Neurosurgery*, Jonathan Borden, a resident in neurosurgery at Tufts, authored a paper with senior author William Shucart that intended to extrapolate the Djindjian cranial scheme to spinal fistulas[10]:

> *We are proposing a classification system that unifies spinal and cranial dural AVFMs and provides a rationale for modes of treatment. Our system is based upon the classification system for cranial dural AVFMs of Djindjian and Merland.*

Borden-Shucart type 1 fistulas linked Djindjian type 1 cranial fistulas to spinal epidural fistulas draining into Batson's epidural plexus. Type 2 fistulas linked Djindjian type 2 cranial fistulas to spinal epidural fistulas with both epidural and intradural venous drainage (commonly seen in spinal epidural fistulas).[11] Type 3 fistulas linked Djindjian type 3 cranial fistulas to typical spinal dural arteriovenous fistulas, often typically referred to as "type 1" spinal AVFs,[12–14] or in Spetzler's scheme, intradural dorsal spinal arteriovenous fistulas.[15] The type 4 designation was dropped as there was no spinal angiographic correlate.

While the Borden-Shucart spinal AV fistula classification scheme did not garner popularity, much of the modern parlance of cranial dAVFs refers to the "Borden classification." This is largely a result of neurosurgical literature promulgation of their scheme as a cranial scheme, failing to recognize the cranial component was Djindjian's from decades before, and the Borden-Shucart scheme is really a spinal arteriovenous fistula classification scheme. Paradoxically, the spinal Borden type 1 or epidural spinal arteriovenous fistula does carry a risk of hemorrhage,[11] unlike type 1 cranial fistulas.[16,17] The spinal Borden type 3 or standard spinal dural arteriovenous fistula is largely associated with venous hypertensive myelopathy and not hemorrhage,[12–14] unlike cranial type 3 fistulas that are associated with venous hypertension and/or hemorrhage.[16,18]

As venous ectasia is a relevant clinical and natural history feature of cranial dAVFs,[16,18] the original Djindjian classification scheme has largely withstood the test of time and remains an ideal dAVF scheme, even though many use the term "Borden classification" for it and omit the relevant distinction of type 3 versus type 4.

## From dural arteriovenous malformation to fistula

Though described in children along variants from pure sinus malformations to fistulae,[19] dAVFs are largely regraded as acquired lesions and as such, modern literature recognizes them as fistulas and not malformations.[16–18] It is noteworthy, however, that the term dural AVM was nearly ubiquitous in the early literature describing these lesions.[20,21] Furthermore, although their radiosurgical management is akin to that for AVMs, their surgical and endovascular management is generally more facile as it is targeted directly to the venous drainage without the need to systematically disconnect arterial pedicles prior to venous disconnection. Instead of harboring a nidus like a true malformation, dAVFs are essentially a localized, "live" piece of dura with arteriovenous shunting within their leaflets. By definition, this makes these lesions unique to the neuraxis, but in addition, by understanding this simple depiction, their arterial and venous anatomy, and their management, becomes quite simplified.

## References

1. Perrini P, Nannini T, DiLorenzo N. Francesco Rizzoli (1809-1880) and the elusive case of Giulia: the description of an "arteriovenous aneurysm passing through the wall of the skull". *Acta Neurochir* 2007;**149**:191–6.
2. Sachs E. *The diagnosis and treatment of brain tumors.* St. Louis, MO: Mosby; 1931.
3. Hayes GJ. External carotid-cavernous sinus fistulas. *J Neurosurg* 1963;**20**:692–700.

4. Djindjian R, Cophignon J, Comoy J, et al. Polymorphisme neuro-radiologique des fistules carotid-caverneuses. *Neurochirurgie* 1968;**14**:881–90.
5. Laine E, Galibert P, Lopez C, et al. Aneurysmses arterioveineux intra-duraux (developpes dans l'epaisseur de la dure-mere) de la fosse posterieure. *Neurochirurgie* 1963;**9**:147–58.
6. Newton TH, Weidner W, Greitz T. Dural arteriovenous malformation in the posterior fossa. *Radiology* 1968;**90**:27–35.
7. Houser OW, Baker HL, Rhoton AL, Okazaki H. Intracranial dural arteriovenous malformations. *Radiology* 1972;**105**:55–64.
8. Djindjian R, Merland JJ. *Superselective angiography of the external carotid artery.* Berlin/Heidelberg/New York: Springer-Verlag; 1978.
9. Cognard C, Gobin YP, Pierot L, et al. Cerebral dural arteriovenous fistulas: clinical and angiographic correlation with a revised classification of venous drainage. *Radiology* 1995;**194**:671–80.
10. Borden JA, Wu JK, Shucart WA. A proposed classification scheme for spinal and cranial dural arteriovenous fistulous malformations and implications for treatment. *J Neurosurg* 1995;**82**:166–79.
11. Huang W, Gross BA, Du R. Spinal extradural arteriovenous fistulae. *J Neurosurg Spine* 2013;**19**:582–90.
12. Ropper AE, Gross BA, Du R. Surgical treatment of type I spinal dural arteriovenous fistulas. *Neurosurg Focus* 2012;**32**, E3.
13. Steinmetz MP, Chow MM, Krishnaney AA, et al. Outcome after the treatment of spinal dural arteriovenous fistulae: a contemporary single-institution series and meta-analysis. *Neurosurgery* 2004;**55**:77–87. discussion 87–78.
14. Saladino A, Atkinson JL, Rabinstein AA, et al. Surgical treatment of spinal dural arteriovenous fistulae: a consecutive series of 154 patients. *Neurosurgery* 2010;**67**:1350–7. discussion 1357–1358.
15. Kim LJ, Spetzler RF. Classification and surgical management of spinal arteriovenous lesions: arteriovenous fistulae and arteriovenous malformations. *Neurosurgery* 2006;**59**:S195–201. discussion S193–113.
16. Gross BA, Albuquerque FC, McDougall CG, et al. A multi-institutional analysis of the untreated course of cerebral dural arteriovenous fistulas. *J Neurosurg* 2018;**129**:1114–9.
17. Satomi J, van Dijk JM, Terbrugge KG, Willinsky RA, Wallace MC. Benign cranial dural arteriovenous fistulas: outcome of conservative management based on the natural history of the lesion. *J Neurosurg* 2002;**97**:767–70.
18. Bulters DO, Mathad N, Culliford D, Millar J, Sparrow OC. The natural history of cranial dural arteriovenous fistulae with cortical venous reflux—the significance of venous ectasia. *Neurosurgery* 2012;**70**:312–9.
19. Gross BA, Akgoz A, Orbach DB. Dural arteriovenous shunts in children. *J Ped Neuroradiol* 2013;**2**:263–8.
20. Awad IA, Little JR, Akarawi WP, Ahl J. Intracranial dural arteriovenous malformations: factors predisposing to an aggressive neurological course. *J Neurosurg* 1990;**72**:839–50.
21. Lasjaunias P, Chiu M, ter Brugge K, Tolia A, Hurth M, Bernstein M. Neurological manifestations of intracranial dural arteriovenous malformations. *J Neurosurg* 1986;**64**:724–30.

# CHAPTER 2

# Anatomical perspective

**Bradley A. Gross**
Department of Neurological Surgery, University of Pittsburgh, Pittsburgh, PA, United States

Unique to the neuraxis, a dural arteriovenous fistula (dAVF) is simply a "live piece of dura" where meningeal arteries shunt into a vein within the dural leaflets. The vein can remain in the dura (sinus/meningeal outflow), or the vein can exit the dura, traverse the subdural space and pressurize/cause venous hypertension of subpial veins. Symptoms from the fistula are predicated on venous outflow. The fistula is named for the associated venous sinus (superior sagittal, transverse-sigmoid, cavernous, or marginal), or in cases where there is not a typical venous sinus, the dural structure (anterior fossa, tentorial, and its subdivisions). A thorough understanding of meningeal anatomy is crucial in the depiction and comprehension of dAVF location and treatment. As an indispensable, timeless reference, Djindjian's *Superselective Angiography of the External Carotid Artery* is highly recommended as a supplement of both historical and clinical value.[1] This chapter reinforces anatomy from two perspectives: (1) evaluating relevant arterial pedicles and (2) evaluating anticipated arterial pedicles for each fistula location.

## Arterial anatomy

A summary of key arterial conduits to the dura is provided in Table 1. It should be noted that any pial artery, including the MCA and ACA can provide transdural branches that traverse the subdural space to a dAVF, particularly in recurrent and highly complex cases (Fig. 1A). A prior craniotomy may serve as substrate for this to occur (Fig. 1B).

## Internal carotid artery branches

The cavernous segment of the internal carotid artery (ICA) provides several relevant dural branches to a variety of dAVFs. Its most proximal branch emanating from the proximal posterior genu, the meningohypophyseal trunk (MHT), can provide early direct supply to cavernous dAVFs (Fig. 2). The MHT trifurcates into the marginal tentorial artery (artery of Bernasconi

*Cerebral Dural Arteriovenous Fistulas*
https://doi.org/10.1016/B978-0-12-819525-3.00007-1

**Table 1** Common parent arteries involved in dAVFs.[a]

| Parent artery | Artery | dAVF supply |
|---|---|---|
| ***ICA*** | | |
| ICA, cavernous segment | MHT, marginal tentorial artery | **Cavernous**, tentorial, T-sig |
| | Clival branches | Marginal sinus |
| | ILT | **Cavernous**, tentorial |
| Ophthalmic artery | Recurrent meningeal | SSS, cavernous, tentorial |
| | Ethmoidal | **AFossa** |
| | Falcine | Anterior SSS |
| ***ECA*** | | |
| Ascending pharyngeal | NMT | **Marginal sinus**, T-sig, torcular, tentorial |
| | NMT, clival branch | Cavernous |
| Occipital | Transosseous | **T-sig**, torcular, tentorial, posterior SSS |
| IMAX | MMA petrosal branches | **Petrosal**, tentorial |
| | MMA petrosquamosal | **T-sig**, tentorial, SSS |
| | MMA sphenoidal | AFossa, cavernous |
| | MMA frontal/parietal | **SSS** |
| | AMA | Cavernous |
| | Sphenopalatine ethmoidal | AFossa |
| | Artery of foramen rotundum | Cavernous |
| ***Posterior circulation*** | | |
| Vertebral (V2) | Anterior meningeal | Marginal sinus |
| Vertebral (V3) | Posterior meningeal | T-sig, torcular, tentorial, posterior SSS |
| Basilar AICA | Subarcuate | Petrosal |
| Basilar SCA | Wollschlaeger and Wollschlaeger | Tentorial |
| PCA | Davidoff and Schecter | Tentorial |

[a] *AFossa*, anterior fossa/ethmoidal; *AICA*, anteroinferior cerebellar artery; *AMA*, accessory meningeal artery; *ECA*, external carotid artery; *ICA*, internal carotid artery; *ILT*, inferolateral trunk; *MHT*, meningohypophyseal trunk; *MMA*, middle meningeal artery; *NMT*, neuromeningeal trunk; *PCA*, posterior cerebral artery; *SCA*, superior cerebellar artery; *SSS*, superior sagittal sinus/convexity; *T-Sig*, transverse-sigmoid.

and Cassinari), the inferior hypophyseal artery and the dorsal meningeal artery. The marginal tentorial artery often supplies cavernous, tentorial, and transverse-sigmoid dAVFs. Adjacent to the MHT, direct ICA clival branches may supply clival and marginal sinus dAVFs (Fig. 3).

The inferolateral trunk (ILT) is a pertinent branch of the mid-horizontal cavernous ICA. It supplies the cavernous sinus and associated cranial nerves

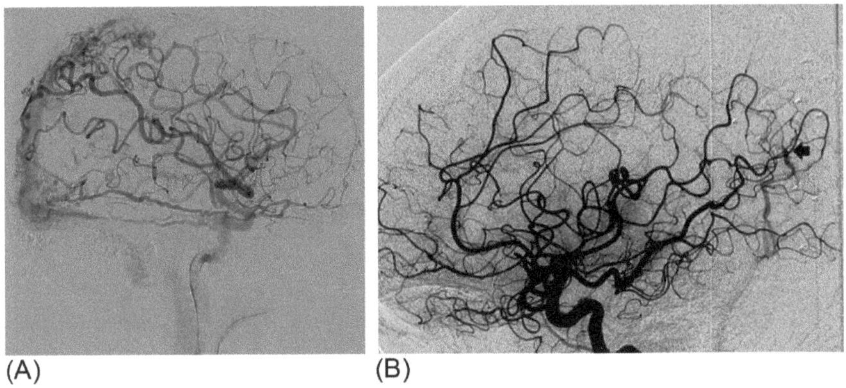


**Fig. 1** Transdural supply from multiple MCA pial branches to a superior sagittal sinus fistula (A); also note the enlarged recurrent meningeal branch of the ophthalmic artery with substantial supply. As another example, following a craniotomy, this dAVF supplied by transdural PCA branches is seen (*arrow*, B).

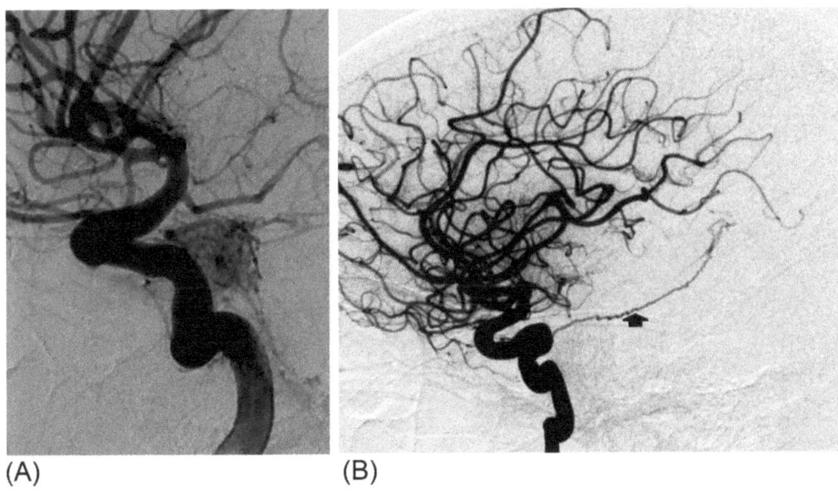


**Fig. 2** A cavernous dAVF supplied by branches of the MHT/proximal marginal tentorial artery (A). A different case of a prominent marginal tentorial artery *(arrow)* to a tentorial/galenic dAVF (B).

and is often involved in cavernous dAVFs (Fig. 4). Involvement in tentorial dAVFs is also seen. Supply to the foramen ovale may anastomose with the ascending accessory meningeal artery and supply to the foramen rotundum may anastomose to the artery of foramen rotundum (Fig. 5).

The ophthalmic artery may supply a variety of dAVFs through recurrent meningeal branches that anastomose with or supplant the middle

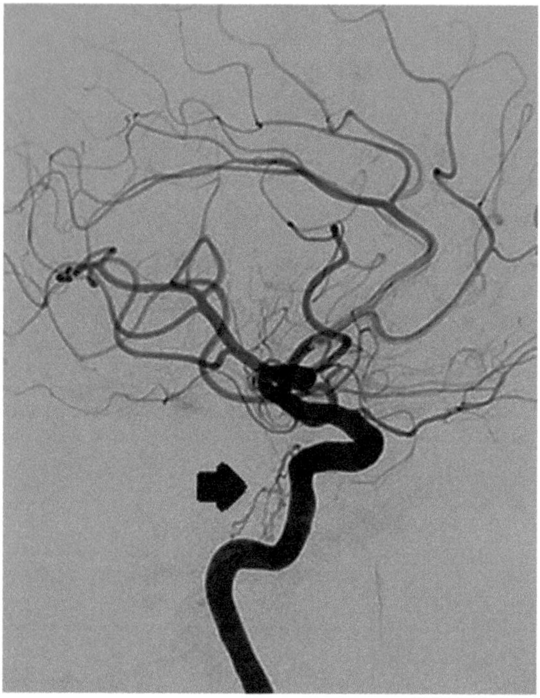

Fig. 3 ICA clival branches to a marginal sinus dAVF.

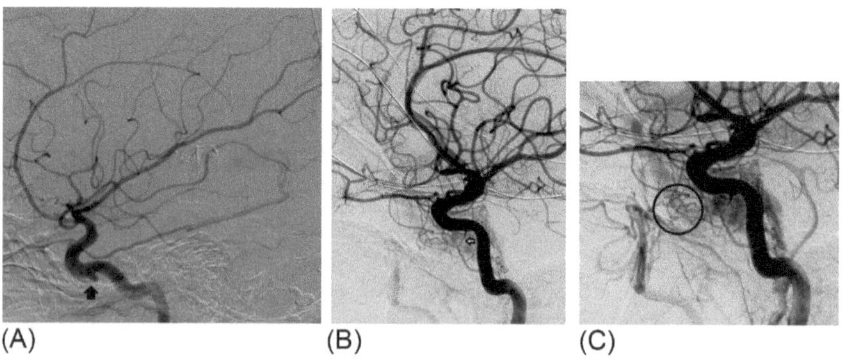


Fig. 4 Prominent ILT (A) involved in a galenic dAVF. ILT supply to a cavernous dAVF (B) with later arterial phase with reflux into the ECA/IMAX showing putative connecting anastomoses to the artery of foramen rotundum (C).

meningeal artery. If replacing the middle meningeal artery, dAVFs along the convexity/superior sagittal sinus may be supplied by the ophthalmic artery (Fig. 6). Recurrent meningeal branches may also supply cavernous dAVFs and even tentorial dAVFs. Distal ethmoidal branches are the primary supply

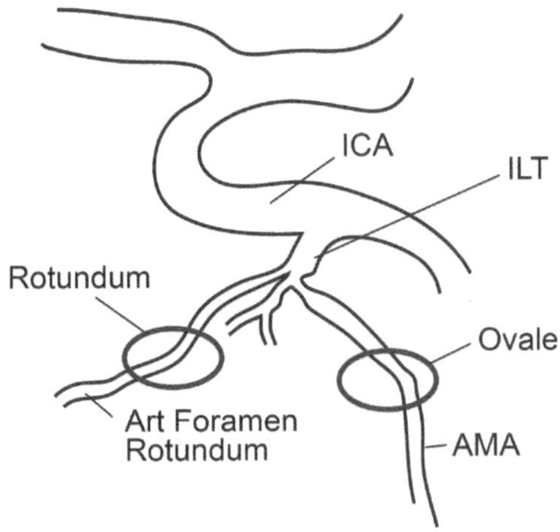


**Fig. 5** Schematic diagram of ILT and anastomoses to the artery of foramen rotundum and accessory meningeal artery.

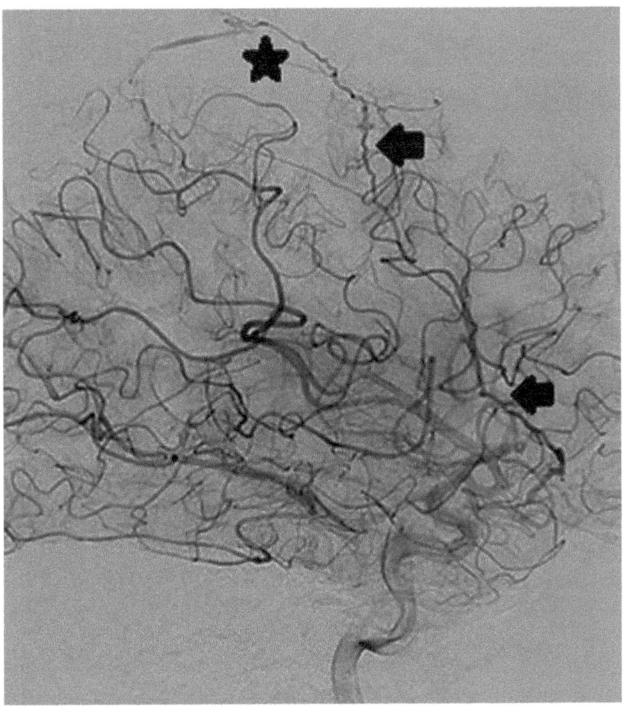

**Fig. 6** Convexity dAVF supplied by the recurrent meningeal branch of the ophthalmic artery *(arrows)*.

to ethmoidal/anterior fossa dAVFs (Fig. 7) and distal falcine branches may be involved in more proximal superior sagittal sinus dAVFs.

## ECA branches

Relevant ECA branches for dAVFs include the ascending pharyngeal artery, occipital artery, and internal maxillary arteries. Transosseous branches of the occipital artery are frequently involved in transverse-sigmoid, posterior superior sagittal sinus/convexity and tentorial dAVFs (Fig. 8). Transosseous branches of the posterior auricular and superficial temporal artery may also be rarely scavenged, though these are frequently overimplicated branches: if suspecting posterior auricular artery supply, then confirm that the branch is not in fact from the ascending pharyngeal or occipital arteries. A detailed, angiographically illustrated discussion of the ascending pharyngeal and internal maxillary artery ensues.

### *Ascending pharyngeal artery*

The ascending pharyngeal artery (APA), which may emanate from the occipital artery, or rarely, the ICA, has early musculospinal branches that may

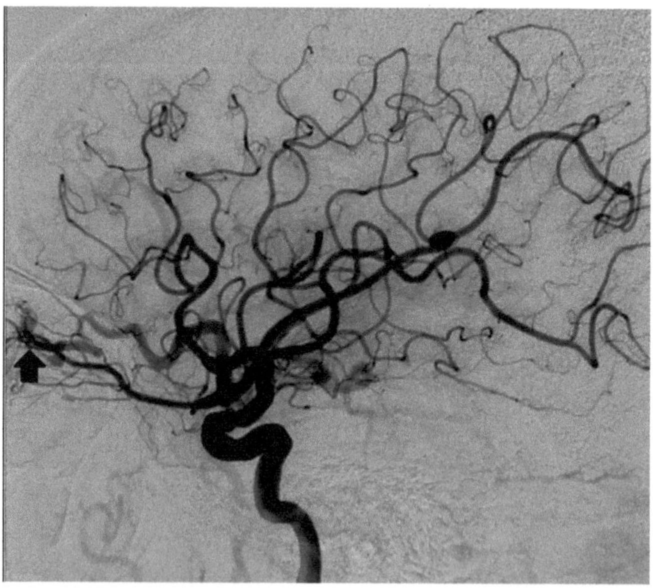

**Fig. 7** Anterior fossa/ethmoidal dAVF with deep venous drainage supplied by ethmoidal branches of the ophthalmic artery.

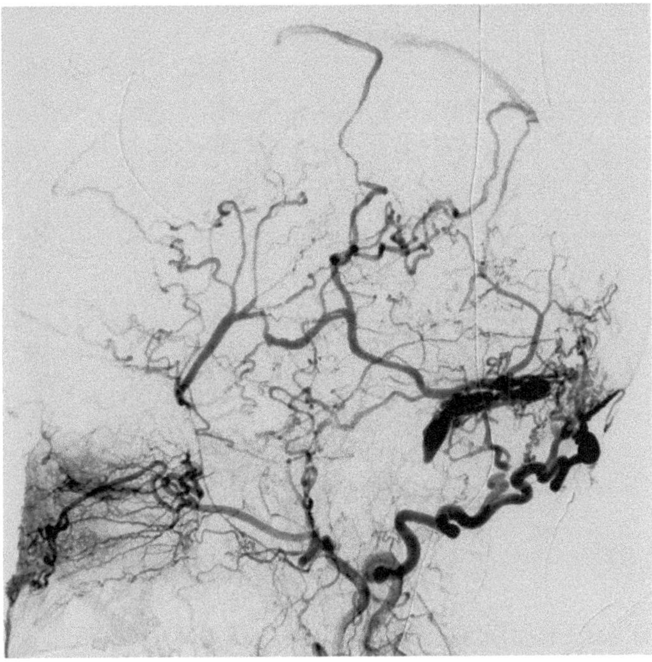

**Fig. 8** Lateral ECA run of a transverse-sigmoid dAVF supplied predominantly via transosseous occipital arteries, and MMA petrosquamosal branches with profound cortical venous reflux.

anastomose with the vertebral artery. The vessel bifurcates into two main trunks: the anteriorly oriented pharyngeal trunk and the posteriorly oriented neuromeningeal trunk (Fig. 9). The inferior tympanic branch typically arises between the two trunks and may anastomose to the petrous ICA via the elusive caroticotympanic artery. The pharyngeal trunk has superior, middle, and inferior pharyngeal branches. The superior pharyngeal artery may anastomose with the ILT. dAVF supply is often via branches of the neuromeningeal trunk (NMT). The NMT provides supply to the lower cranial nerves via its jugular and hypoglossal branches. A terminal clival branch may anastomose with the ICA or alternatively supply dAVFs along the clivus or cavernous dAVFs (Fig. 10). Most commonly, branches of the NMT are the backbone supply to marginal sinus dAVFs (Fig. 11) and may also commonly supply transverse-sigmoid dAVFs. Lower cranial nerve supply makes the NMT a suboptimal transarterial conduit for embolization unless the venous pouch can be accessed.

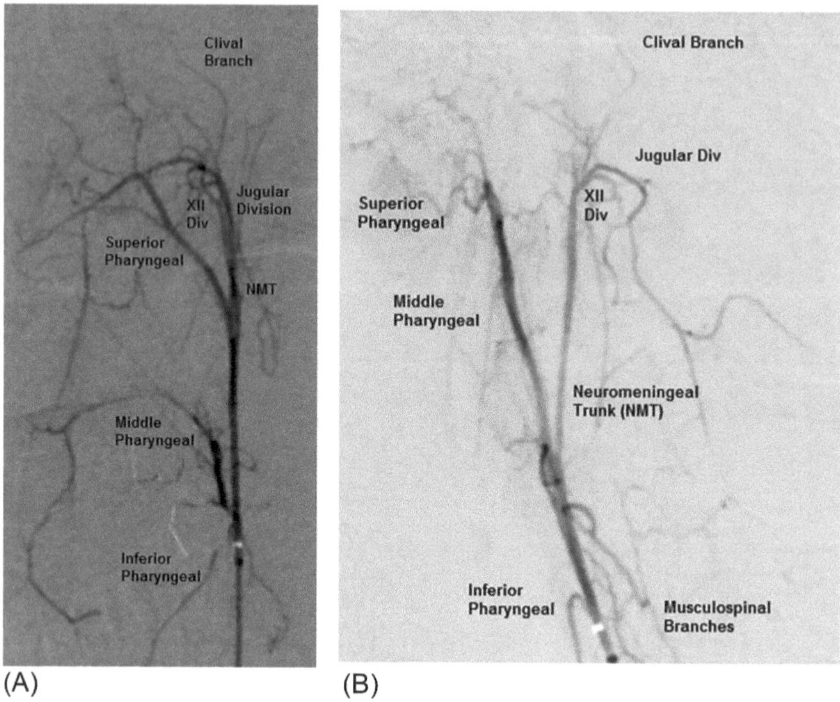


**Fig. 9** AP (A) and lateral (B) views of a superselective ascending pharyngeal artery injection.

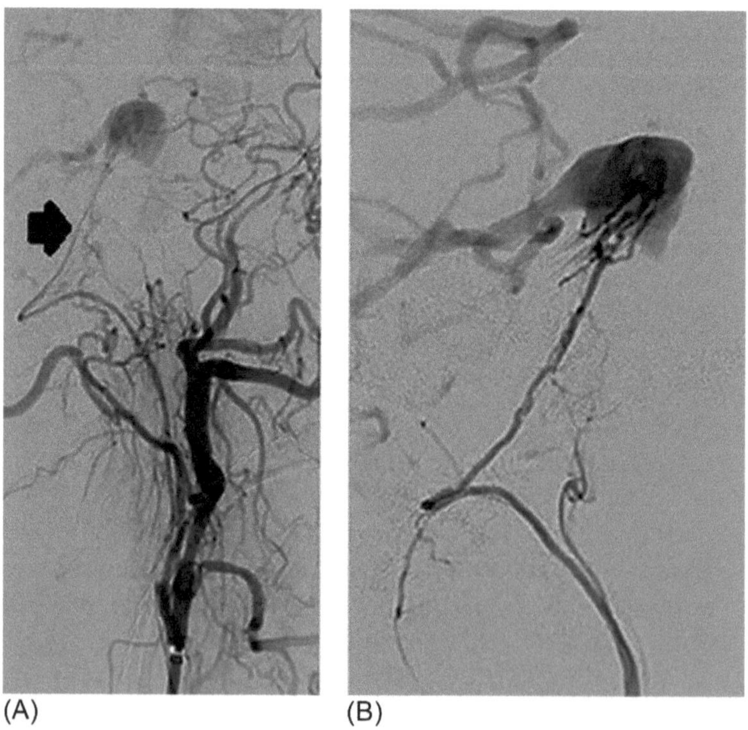


**Fig. 10** Clival branch of the neuromeningeal trunk supplying a cavernous dAVF (A, *arrow*; B, *magnified view*).

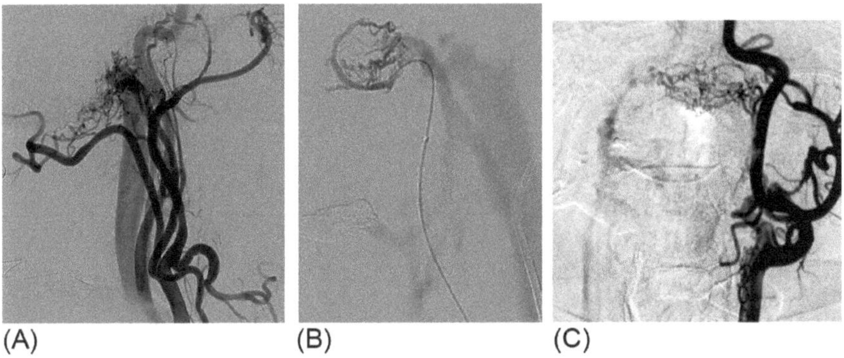


**Fig. 11** Typical Type I marginal sinus fistula supplied by primarily the ascending pharyngeal artery (A), with a superselective ascending pharyngeal injection (B). Commonly, marginal sinus fistulas may recruit the contralateral ascending pharyngeal artery (C).

### *Internal maxillary artery*

A thorough understanding of relevant internal maxillary artery (IMAX) anatomy holds innumerable applications and is particularly relevant for dAVFs (Fig. 12). Oft-implicated branches are the middle meningeal artery (MMA), involved in innumerable dAVF subtypes, the accessory meningeal artery and artery of foramen rotundum, involved in cavernous fistulas, and ethmoidal branches of sphenopalatine arteries involved in anterior fossa/ethmoidal fistulas.

The IMAX can be divided into three segments. The first segment, the mandibular segment gives rise to deep auricular, anterior tympanic, middle meningeal, accessory meningeal, and inferior alveolar arteries. The deep auricular and anterior tympanic branches are small, posteriorly oriented branches to the ear. The inferior alveolar artery is a constant, relatively large, inferiorly oriented artery that terminates in the mental artery.

The MMA emanates from the first segment of the internal maxillary artery with a classic direct, superior course that angulates 90 degrees just above foramen spinosum. On the anteroposterior projection, this angulation results in a lateral course while on the lateral projection, it results in an anteriorly oriented course along the middle fossa floor. The first potential relevant branch of the MMA is the accessory meningeal artery (AMA) that may arise along the proximal aspect of its initial ascending, extracranial segment.

Once intracranial, the MMA provides small, posteriorly oriented petrosal branches (Fig. 13A). Entry of either liquid embolics or particles into petrosal branches should be avoided, as it may result in facial palsy or

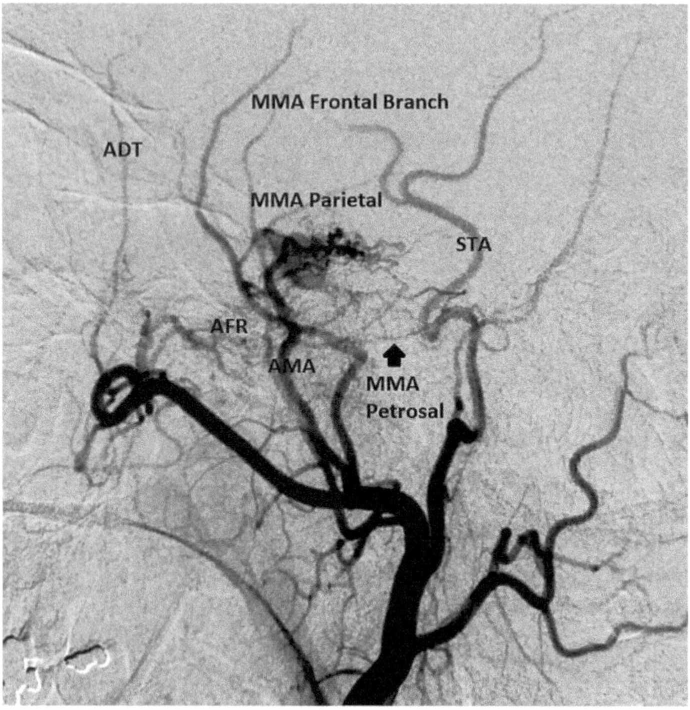


**Fig. 12** Highlighted IMAX branches in a patient with a cavernous dAVF (*ADT*, anterior deep temporal artery; *AFR*, artery of foramen rotundum; *AMA*, accessory meningeal artery; *MMA*, middle meningeal artery; *STA*, superficial temporal artery).

dysfunction of the greater superficial petrosal nerve (lack of tears and loss of taste). Petrosal branches almost invariably supply petrosal dAVFs and can also supply tentorial dAVFs. An often larger, subsequent early posteriorly oriented branch is the petrosquamosal branch that is involved a variety of dAVFs and is often an excellent transarterial conduit (Fig. 13B). After coursing anteriorly and superiorly, the MMA gives rise to anteromedially oriented sphenoidal branches that have variable degrees of orbital supply and may also supply fistulas along the sphenoid ridge, cavernous sinus, or ethmoidal/anterior fossa dAVFs (Fig. 13C and D). In the absence of an ophthalmic artery from the ICA, sphenoidal branches course through the superior orbital fissure as meningo-ophthalmic to supply the choroidal blush (Fig. 14). Alternatively, supplying the lateral orbit and lacrimal gland, these branches course through the foramen of Hyrtl as meningolacrimal branches that may still anastomose with the ophthalmic artery. Following the takeoff of sphenoidal branches, the MMA then bifurcates into frontal and parietal branches; however, parietal branches may alternatively emanate more

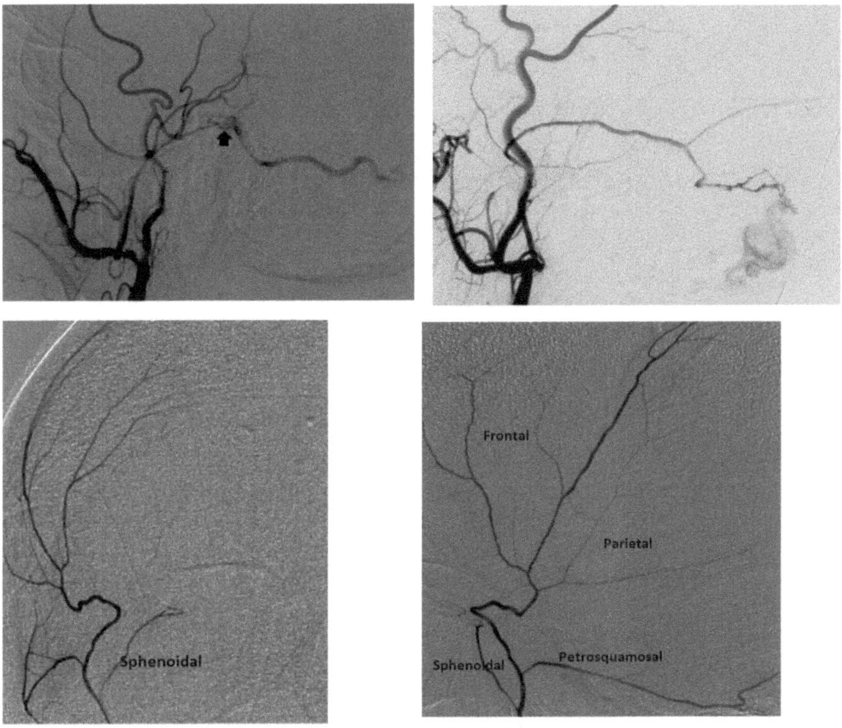


**Fig. 13** MMA branches. Petrosal branch (A) supplying a Type III dAVF. Petrosquamosal branch supplying a Type IV dAVF (B). Slightly distal MMA superselective AP (C) and lateral (D) views highlighting relevant, labeled branches.

proximally from the petrosquamosal branch, or in lieu of the petrosquamosal branch. These almost invariably supply convexity/superior sagittal sinus dAVFs (Fig. 15).

The accessory meningeal artery often emanates from the proximal MMA or adjacent to the MMA ostium from the IMAX itself. It has extracranial supply to pterygoid muscles but of relevance will pass through foramen ovale to supply the fifth nerve and cavernous sinus. As such, it can certainly be involved in cavernous dAVFs (Fig. 16). It also may anastomose with the ILT.

The second segment of the IMAX is the pterygoid segment, giving rise to deep temporal arteries, buccal, masseteric, and pterygoid branches. The third portion of the IMAX, the pterygopalatine segment, gives rise to the posterior superior alveolar artery, greater palatine artery, sphenopalatine arteries, and the infraorbital artery. The terminal course via the infraorbital artery passes through the inferior orbital fissure and then out the orbit through the infraorbital foramen with terminal branches including the anterior superior alveolar artery.

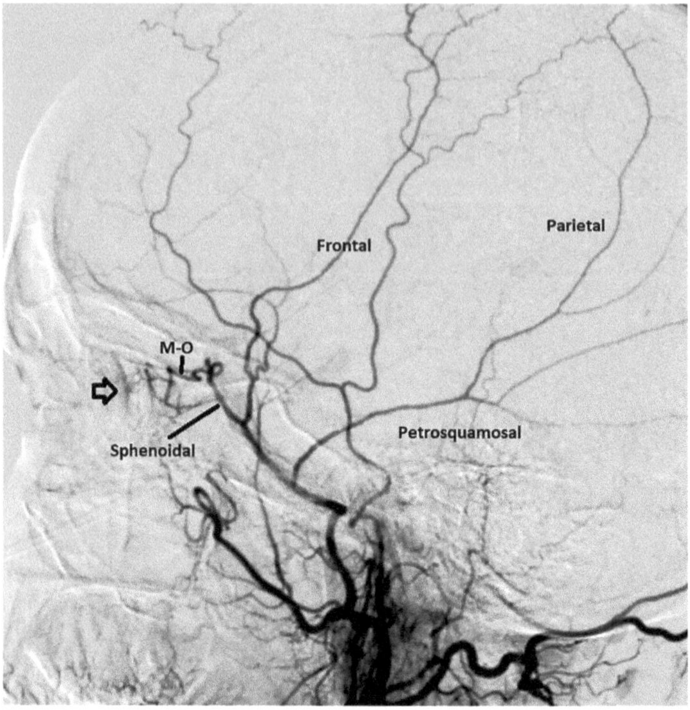


**Fig. 14** This patient did not have a radiographically demonstrable ophthalmic artery from the ICA. Note the meningo-ophthalmic artery (MO) emanating from an enlarged sphenoidal branch.

Sphenopalatine arteries give rise to small ethmoidal branches that may supply ethmoidal dAVFs, often not amenable to transarterial approaches through these tiny conduits. In selected cases, a recurrent/posteriorly oriented branch from the proximal third segment, the artery of foramen rotundum, courses through the foramen rotundum to supply the fifth nerve, cavernous sinus, and potentially anastomose with the ILT. As such, it can often be involved in cavernous dAVFs (Fig. 17).

## Vertebral and posterior circulation branches

The anterior meningeal branch of the distal V2 may be involved in marginal sinus dAVFs as the vessel supplies ventral/clival dura (Fig. 18). The posterior meningeal artery is involved in a variety of dAVFs including posterior convexity, transverse-sigmoid, torcular, and tentorial dAVFs (Fig. 19). A subarcuate branch of the anteroinferior cerebellar artery (AICA) may supply lateral tentorial and/or petrosal dura (Fig. 20). A superior cerebellar artery (SCA) dural branch, the artery of Wollschlaeger and Wollschlaeger, may supply tentorial dAVFs as can a better known PCA dural branch, the artery of Davidoff and Schecter (Fig. 21).

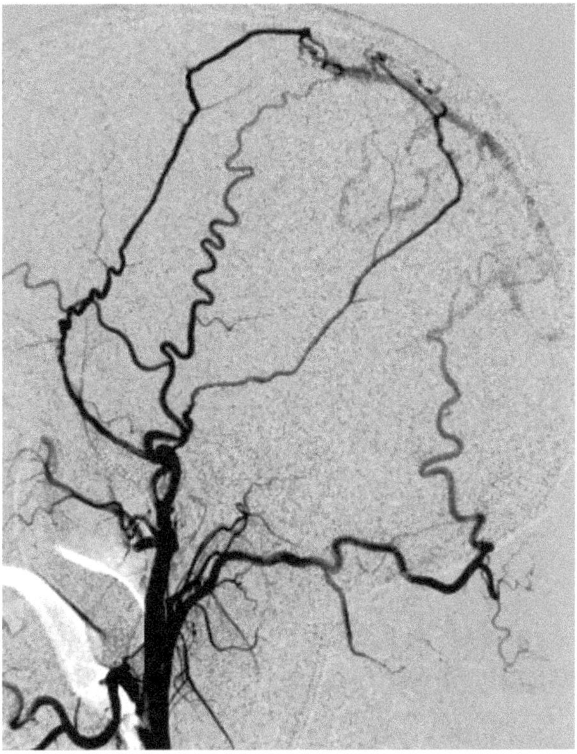

**Fig. 15** Convexity/superior sagittal sinus Type III dAVF supplied by MMA frontal and parietal branches.

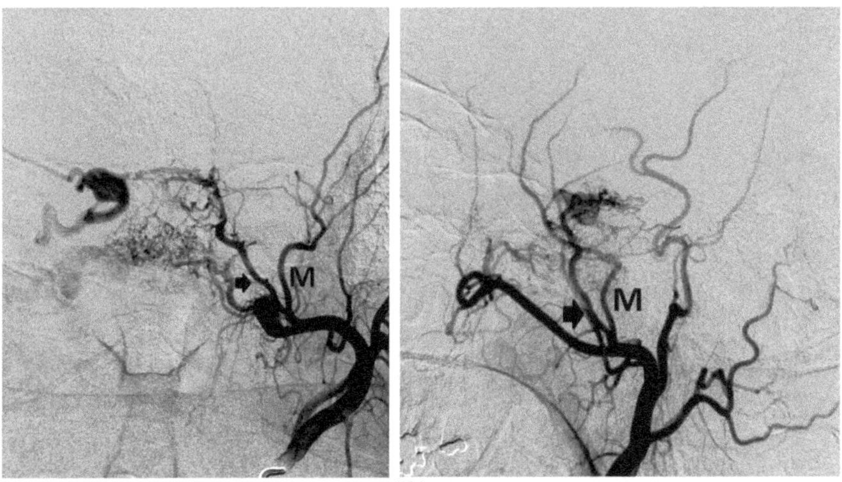


(A) (B)

**Fig. 16** AP (A) and lateral (B) distal ECA runs demonstrating the accessory meningeal artery supply *(arrow)* to a contralateral Type II cavernous dAVF; M denotes the middle meningeal artery.

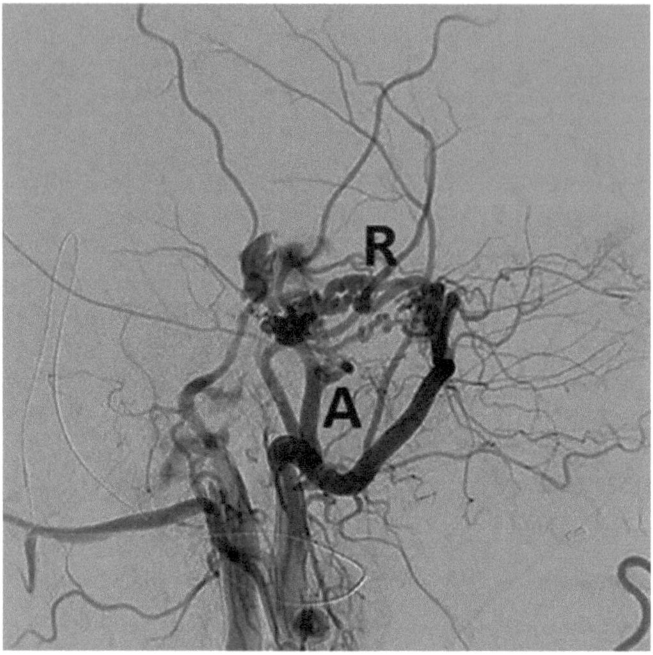


**Fig. 17** Lateral distal ECA run demonstrates accessory meningeal (A) and artery of foramen rotundum (R) supply to a cavernous dAVF.

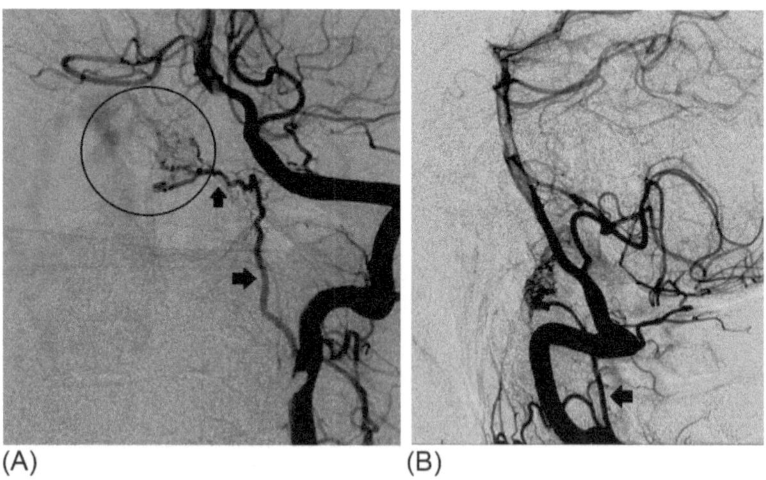


**Fig. 18** AP (A) and lateral (B) left vertebral runs demonstrating anterior meningeal artery supply to a marginal sinus dAVF.

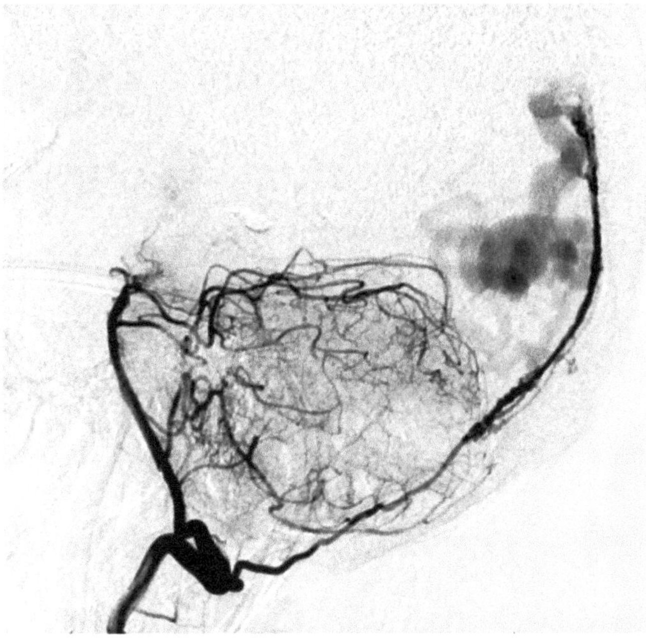

**Fig. 19** Posterior meningeal artery supply to a Type IV posterior convexity/superior sagittal sinus dAVF.

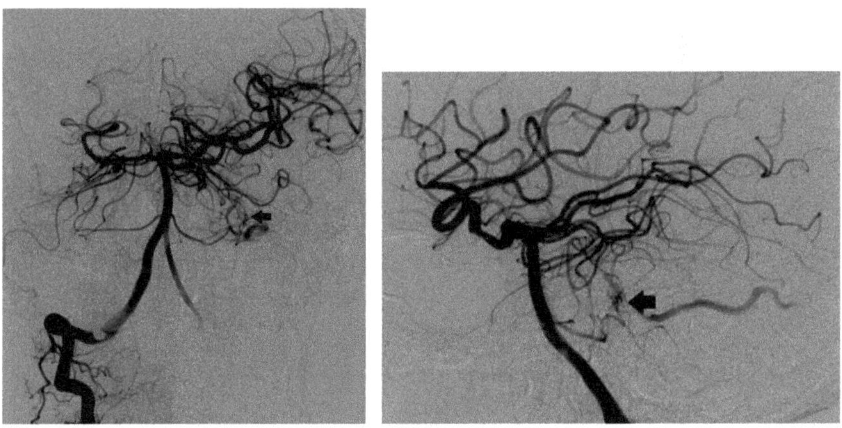

(A) (B)

**Fig. 20** AP (A) and lateral (B) right vertebral runs demonstrating the left AICA subarcuate dural artery branch supplying a petrosal dAVF.

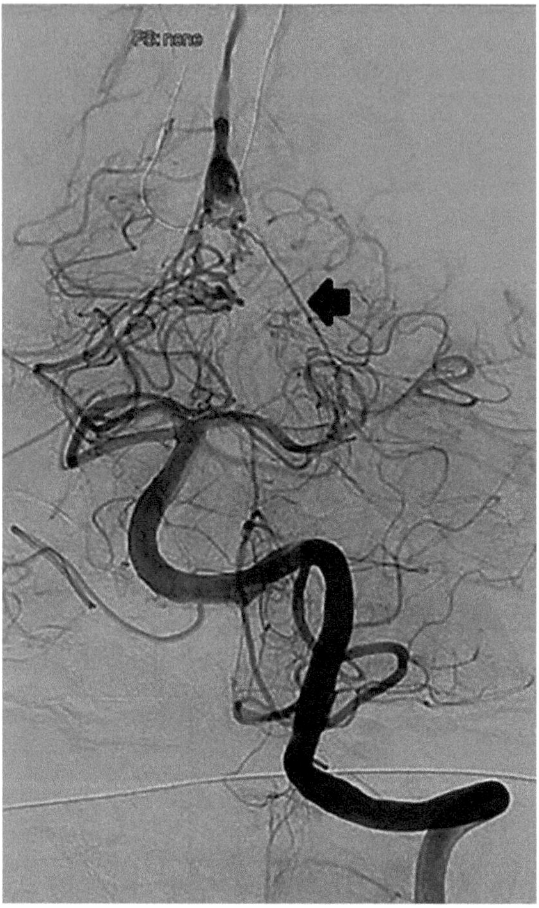

Fig. 21 The artery of Davidoff and Schecter *(arrow)* supplies a tentorial dAVF.

## Fistula location/venous-based anatomy

Fistula location is delineated by a major associated sinus, and, if none is generally present, the dural vicinity (anterior fossa or tentorial) (Table 2). This classification delineates cavernous, superior sagittal sinus/convexity, transverse-sigmoid, and marginal sinuses as major sinus-based fistula locations.

Cavernous sinus dAVFs are best known for their symptoms occurring as a result of reflux into the superior ophthalmic vein and consequent ocular venous hypertension. In a recent large series, they were the only fistula location type with a slight female sex predilection.[2] The oft-cited Barrow

**Table 2** Typical arterial supply and venous outflow pattern by common dAVF locations.[a]

| Location | Potential arterial supply | Venous outflow |
|---|---|---|
| Cavernous | Marginal tentorial<br>ILT<br>AMA<br>Artery foramen rotundum<br>MMA sphenoidal<br>NMT clival | Approximately one-third with cortical venous reflux |
| SSS | MMA<br>Ophthalmic falcine branch<br>Occipital<br>Posterior meningeal artery | Nearly all with cortical venous reflux |
| T-sig | MMA<br>Occipital<br>Marginal tentorial<br>NMT<br>Posterior meningeal | Approximately one-half with cortical venous reflux |
| Marginal sinus | NMT<br>Occipital<br>Anterior meningeal<br>ICA clival | Few (less than one-fourth) with cortical venous reflux<br>May reflux into cavernous sinus/SOV |
| AFossa | Ophthalmic ethmoidal<br>Sphenopalatine ethmoidal<br>MMA sphenoidal | Nearly all with cortical venous reflux<br>Highest proportion with venous ectasia |
| Tentorial | MMA<br>Marginal tentorial<br>ILT<br>NMT<br>Occipital<br>Posterior meningeal<br>ADS | Nearly all with cortical venous reflux |
| Petrosal | MMA<br>Marginal tentorial<br>ILT<br>Subarcuate | Nearly all with cortical venous reflux |

[a] *AFossa*, anterior fossa/ethmoidal; *AMA*, accessory meningeal artery; *ICA*, internal carotid artery; *ILT*, inferolateral trunk; *MHT*, meningohypophyseal trunk; *MMA*, middle meningeal artery; *NMT*, neuromeningeal trunk; *SSS*, superior sagittal sinus/convexity.

classification stratifies carotid-cavernous fistulae into four types.[3] Type A fistulae are direct between the ICA cavernous segment and cavernous sinus occurring as a result of traumatic tear/dissection, aneurysm rupture, or idiopathic/spontaneous physiology. This pathology is entirely different from dAVFs. Type B and C carotid-cavernous fistulae refer to indirect or dAVF supplied exclusively by external and internal carotid branches, respectively. Type D fistulae refer to dAVFs supplied by both internal and external branches representing what is by far the most common dAVF type (~90%). Supply is often seen from the MHT and ILT from the ICA (Figs. 2A and 4B), and the accessory meningeal artery and artery of foramen rotundum from the ECA (Fig. 17). Variably, MMA sphenoidal and APA NMT clival branches (Fig. 10) may also supply these dAVFs. According to a recent series, in approximately one-third of cases cortical the venous reflux is seen (Fig. 16).

Superior sagittal sinus/convexity dAVF are predominantly paramedian and typically drain directly into adjacent convexity cortical veins. Supply is nearly ubiquitously via the MMA (Fig. 15). Anterior convexity lesions may recruit ophthalmic falcine branch supply while posterior convexity lesions may recruit transosseous occipital artery and posterior meningeal artery supply (Fig. 19).

Transverse-sigmoid dAVFs are the "classic" and most common dAVF type supplied by MMA, occipital and often the marginal tentorial branch of the ICA. NMT and posterior meningeal supply may also be recruited. They may present with tinnitus, though hemorrhage is also possible as approximately half may be associated with cortical venous reflux (Fig. 8).

Marginal sinus dAVFs harbor multiple synonymous location terms in the literature including hypoglossal canal and jugular foramen. As other dAVFs are generally not named for associated foramina but rather a major venous sinus, it seems most consistent to refer to these as marginal sinus dAVFs. Cardinal supply is via the APA NMT branch (Fig. 11). Occipital, anterior meningeal (Fig. 18), and ICA clival branch supply (Fig. 3) may also be seen. Venous outflow is typically antegrade, and as such, these dAVFs often present with pulsatile tinnitus. A small proportion may reflux into the cavernous sinus and SOV with consequent ocular venous hypertension and a smaller proportion may reflux into cortical veins (Fig. 22).

Anterior fossa/ethmoidal dAVFs represent the anterior extent of convexity/superior sagittal sinus dAVFs in the true absence of an associated venous sinus (Fig. 7). As such, they essentially ubiquitously harbor cortical venous reflux and thus present either incidentally or as a result of hemorrhage. Symptomatic venous hypertension in lieu of hemorrhage (nonhemorrhagic neurologic deficit) is rare, likely as a result of the dAVF location. This dAVF subclass harbor the highest proportion of lesions with venous

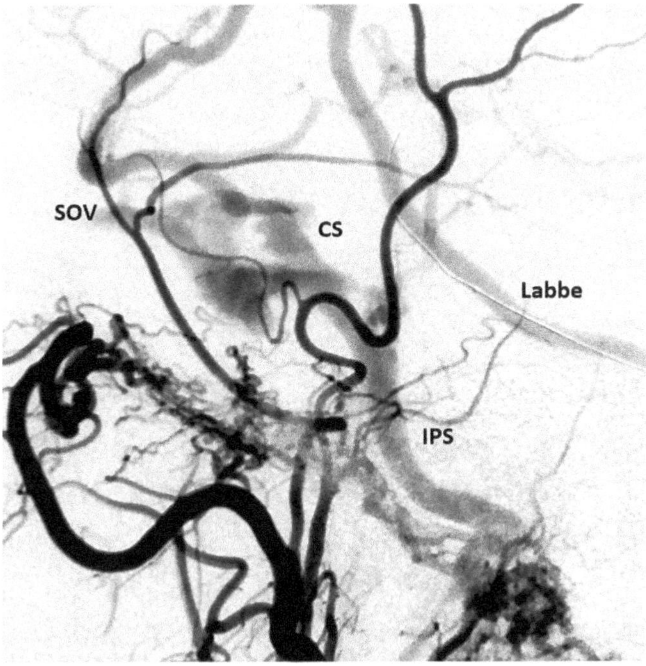


**Fig. 22** Lateral ECA run of a marginal sinus fistula with reflux into inferior petrosal sinus (IPS), cavernous sinus (CS), superior ophthalmic vein (SOV), and cortical veins (via the deep middle cerebral vein); the patient presented with chemosis and proptosis.

ectasia.[2] Arterial supply is invariably via ethmoidal branches of the ophthalmic artery with a variable degree of supply from sphenopalatine ethmoidal branches and less so, sphenoidal MMA branches.

Tentorial dAVFs represent a relatively broad subgroup of lesions occurring along the tentorium. As such, they nearly ubiquitously harbor cortical venous reflux. Supply is via MMA, marginal tentorial, and ILT branches. Variably, NMT, transosseous occipital, posterior meningeal artery, and SCA or PCA dural branch supply can be seen (Fig. 21). To facilitate surgical approach selection, they can be stratified into galenic, straight sinus, and torcular subtypes.

Petrosal dAVFs denote a separate, infratentorial subtype of dAVF, supplied by petrosal branches of the MMA and often the marginal tentorial artery (Fig. 23). ILT and AICA dural branch supply (Fig. 20) can also be recruited and drainage is typically into cortical veins.

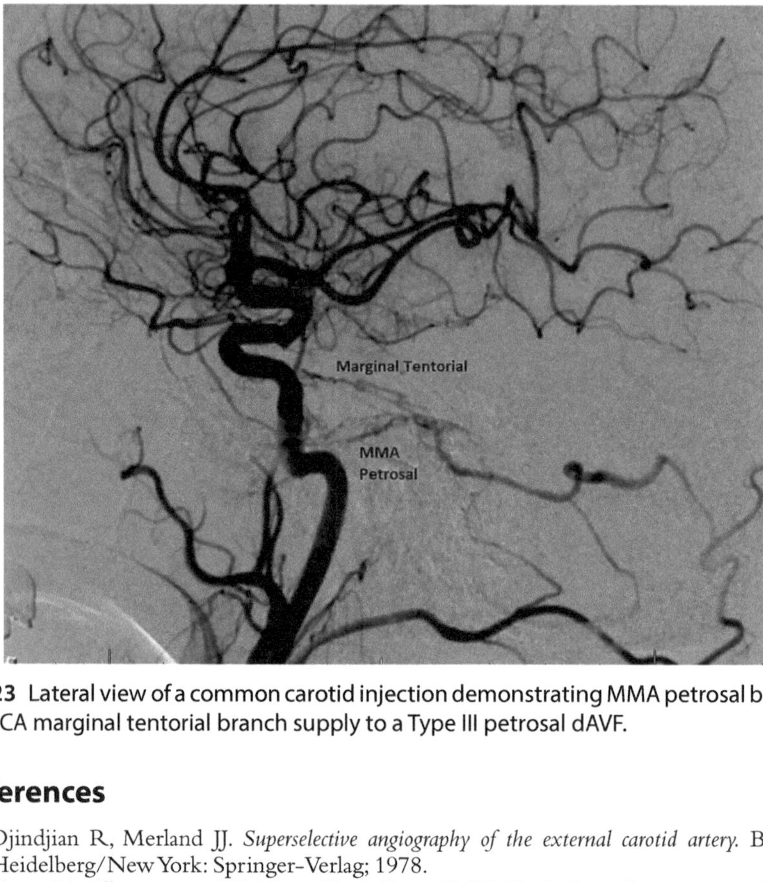


**Fig. 23** Lateral view of a common carotid injection demonstrating MMA petrosal branch and ICA marginal tentorial branch supply to a Type III petrosal dAVF.

## References

1. Djindjian R, Merland JJ. *Superselective angiography of the external carotid artery.* Berlin/Heidelberg/New York: Springer-Verlag; 1978.
2. Gross BA, Albuquerque FC, Moon K, McDougall CG. Evolution of treatment and a detailed analysis of occlusion, recurrence and clinical outcomes in an endovascular library of 260 dural arteriovenous fistulas. *J Neurosurg* 2017;**126**:1884–93.
3. Barrow DL, Spector RH, Braun IF, Landman JA, Tindall SC, Tindall GT. Classification and treatment of spontaneous carotid-cavernous sinus fistulas. *J Neurosurg* 1985;**62**:248–56.

# CHAPTER 3

# Neuroimaging and diagnosis

**Bradley A. Gross**
Department of Neurological Surgery, University of Pittsburgh, Pittsburgh, PA, United States

Dural arteriovenous fistulas (dAVFs) may present "typically" with symptoms such as pulsatile tinnitus or neurological deficits secondary to hemorrhage in association with distended veins. They may alternatively masquerade and serve as an initially elusive etiologic diagnosis. This chapter distills typical/anticipated presentation modalities, presents masquerading features, and highlights of more subtle clinical symptomatology. A focused discussion on pertinent neuroimaging technique and findings then ensues. A summary by typical dAVF location is provided in the Table 1.

## Clinical presentation

Typical clinical presentations for dAVFs include pulsatile tinnitus, ophthalmologic symptoms for cavernous lesions, and hemorrhage for those with cortical venous drainage. Pulsatile tinnitus is most often associated with transverse-sigmoid lesions or marginal sinus fistulas, though may be associated with fistulas in other locations at the skull base.[1,2] Fistulas presenting with pulsatile tinnitus may or may not harbor cortical venous drainage.

Ophthalmologic symptomatology is most associated with cavernous dAVFs owing to reflux into the superior ophthalmic vein, resulting in chemosis, proptosis, and elevated intraocular pressure. Ophthalmoparesis may also occur. Importantly, marginal sinus lesions with reflux into the inferior petrosal sinus and cavernous sinus may also present with ophthalmologic symptomatology (Fig. 1).[1]

Any dAVF with cortical venous reflux may present with hemorrhage; however, only a minority of marginal sinus and cavernous sinus dAVFs have cortical venous reflux.[1,2] Anterior fossa/ethmoidal dAVFs typically present with hemorrhage or are otherwise found incidentally.[1,3] It is important to emphasize that hemorrhage patterns may mimic arteriovenous malformations (AVMs) (intraparenchymal), but fairly uniquely may also include subdural hemorrhage, which should prompt suspicion for a dAVF.

The most intriguing presentation modality that often spawns a broad differential is symptomatic venous hypertension. Patients may present with

*Cerebral Dural Arteriovenous Fistulas*
https://doi.org/10.1016/B978-0-12-819525-3.00005-8

**Table 1** Typical clinical features and key neuroimaging findings for common dAVF types.[a]

| Location | Clinical features | Neuroimaging findings |
|---|---|---|
| Cavernous | Chemosis, proptosis<br>Ophthalmoparesis | Dilated SOV<br>Preponderant vessels in region of cavernous sinus |
| SSS | Focal deficit/seizures from venous HTN<br>Convexity hemorrhage | "Corkscrew veins" at the convexity<br>MRI with unexplained flair signal at the convexity |
| T-sig | Pulsatile tinnitus<br>Focal deficit/seizures from venous HTN | Preponderant vessels/enlarged occipital branches in occipital/ suboccipital region<br>Noncontrast CT with large transosseous channels for occipital branches<br>Inferior temporal "corkscrew" veins if venous reflux |
| Marginal sinus | Pulsatile tinnitus<br>Cavernous Sx if IPS reflux<br>Posterior fossa hemorrhage if CVR | May need common injection to see ascending pharyngeal best<br>Preponderant vessels in region of foramen magnum |
| AFossa | Incidental or hemorrhage<br>Rarely symptomatic NHND given location | Unexplained distended vessels along anterior fossa floor |
| Tentorial/ petrosal | Unusual Sx from venous reflux<br>Hemorrhage | "Corkscrew" veins (often have CVR) |

[a] *AFossa*, anterior fossa; *CVR*, cortical venous reflux; *HTN*, hypertension; *IPS*, inferior petrosal sinus; *NHND*, nonhemorrhagic neurological deficit; *SOV*, superior ophthalmic vein; *Sx*, symptoms.

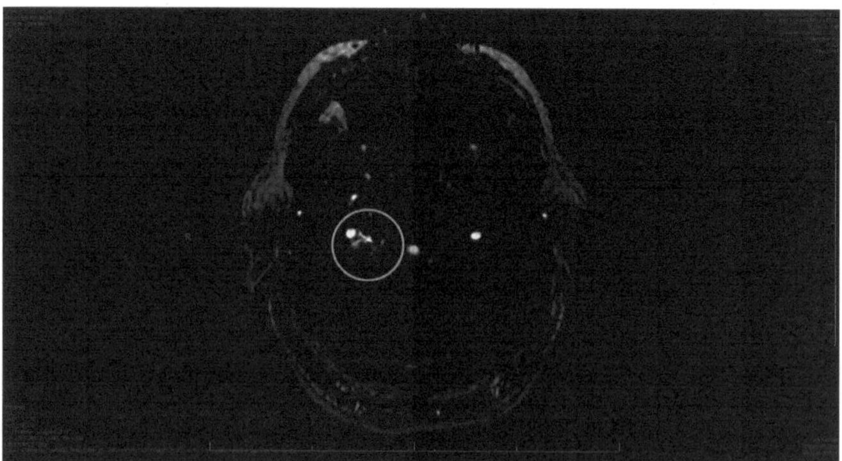

(A)

**Fig. 1** See the figure legend on opposite page

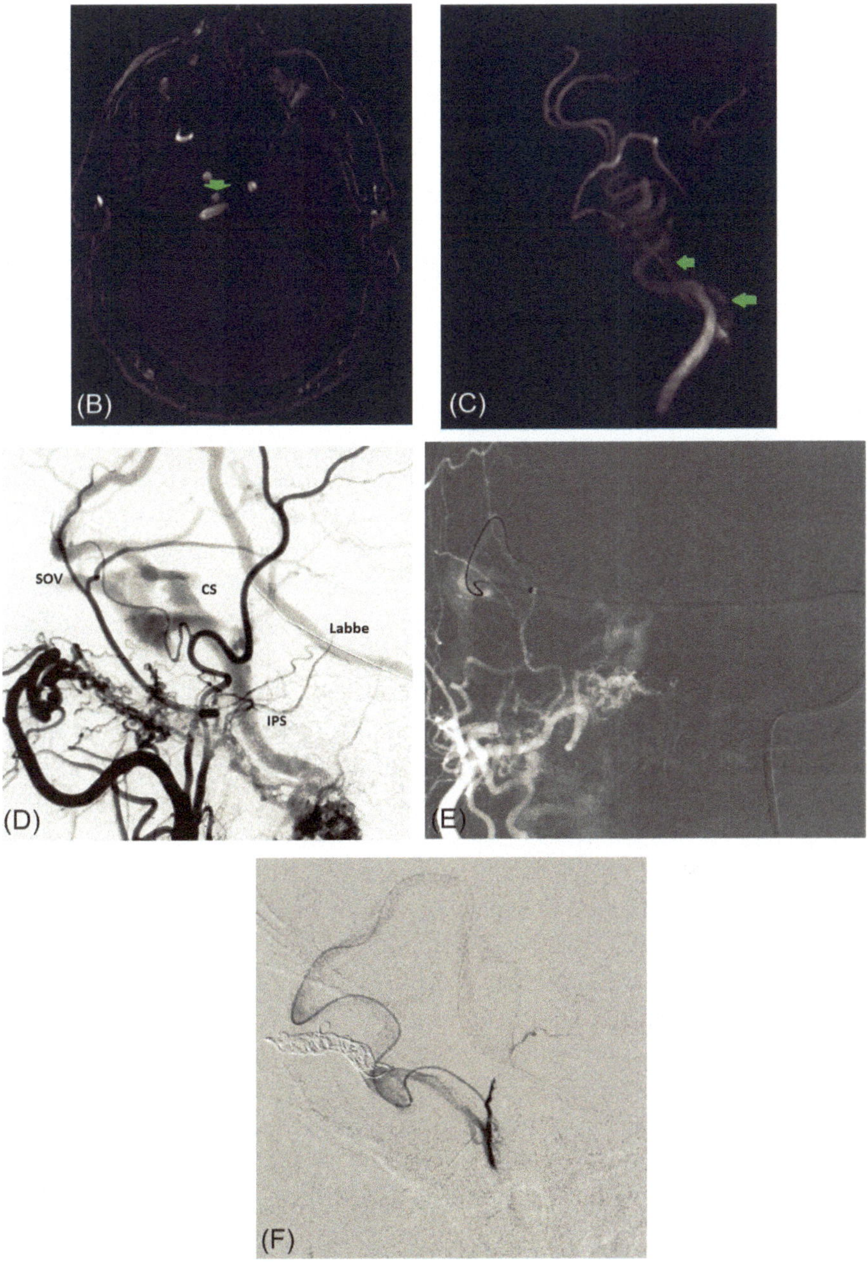


**Fig. 1** Marginal sinus dAVF with reflux into the IPS and cavernous sinus as well as cortical veins presenting with ipsilateral chemosis and proptosis. MRA demonstrates a preponderance of vessels at the right jugular vein (A), with a visible inferior petrosal sinus (*arrows*, B and C). Lateral view of an external carotid run with labeled refluxing veins (D). The fistula was treated via a contralateral transvenous coiling approach (E, F).

nonhemorrhagic neurological deficits (NHND), including a variety of symptoms such as focal deficits, seizures, global neurological dysfunction or dementia, and even Parkinsonism.[4] In one series of 389 dAVFs, six patients presented with progressive dementia (1.4%): four had superior sagittal sinus dAVFs, one had a tentorial dAVF, and one had a transverse-sigmoid dAVF.[5] Endovascular embolization resulted in reversal of dementia-related symptoms.

## Neuroimaging and diagnosis

Digital subtraction angiography is the mandatory gold standard for the diagnosis of dAVFs. Nevertheless, important cues should always be sought on noninvasive modalities such as CT angiography (CTA) and MRI. Studies of the sensitivity and specificity of these modalities identifying dAVFs may be hampered by bias. Only those that evaluate a fully blinded prospective evaluation of the sensitivity of these modalities would prove of true clinical value. A retrospective review of a CTA or MRI with a known dAVF diagnosis will undoubtedly inflate the sensitivity and specificity of the test.

In one study, the sensitivity and specificity of five CTA features were evaluated in a cohort of patients with dAVFs compared with controls.[6] Asymmetrically, increased arterial feeders had a sensitivity of 86% and specificity of 100%, though this can be somewhat subjective and limited by bony anatomy. Asymmetrically increased venous collaterals had a sensitivity of 42% and specificity of 71%. Transcalvarial channels had a sensitivity of 29% and specificity of 86%. A "shaggy sinus or tentorium" had sensitivity of 42% and specificity of 100%. Prominent cortical venous drainage had a sensitivity of 29% and specificity of 71%. These results underscore that CTA is an imperfect modality in the detection of these lesions; as a screening tool, a highly sensitive measure should be sought. Potentially confounded by bias/retrospective analysis of cases in incorporated studies, a systematic review of the literature reported a sensitivity and specificity of 0.90 and 0.94 for MRI, respectively. Sensitivity and specificity for CT was 0.80 and 0.87, respectively.[7] Most neuroimaging studies however are also limited by relatively small numbers.

On CTA, a preponderance of vessels adjacent to a well-opacified transverse-sigmoid system (Fig. 2) or cavernous sinus (Fig. 3) may be a cue for an underlying dAVF. Enlarged vessels along the anterior fossa may be an indicator of an anterior fossa dAVF (Fig. 4), particularly if there is a variceal component. However, telltale sign of a dAVF with cortical venous reflux is "corkscrew veins" (Figs. 5 and 6). Unlike AVMs that will often harbor

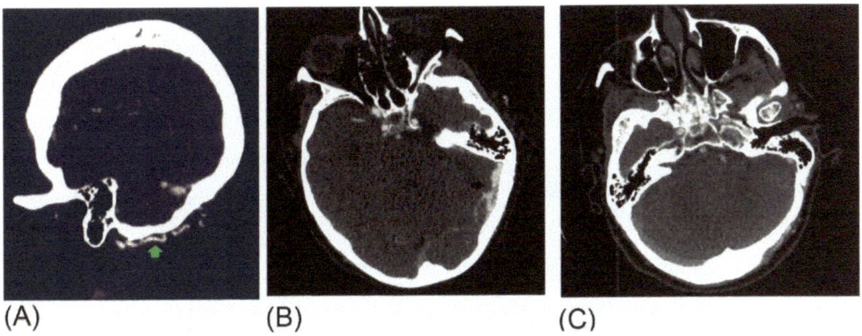


**Fig. 2** Typical transverse-sigmoid sinus dAVF appearance on CTA. Sagittal cut demonstrates a dilated occipital artery (*arrow*, A). Axial cut demonstrates a well-opacified (early) left transverse-sigmoid system (B) in contrast to the normal right-sided system (C).

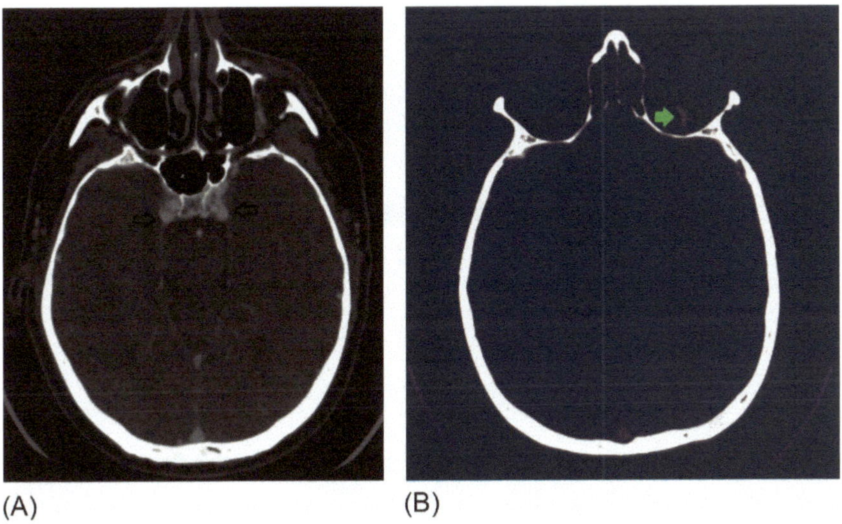


**Fig. 3** Typical cavernous dAVF on CTA. Well-visualized cavernous sinuses bilaterally (*arrows*, A, axial cut) is a sensitive but not entirely specific finding. A dilated superior ophthalmic vein is more telltale (*arrow*, B, CTA axial cut).

enlarged arterial feeders and/or an identifiable nidal area, dAVFs may simply appear on CTA as refluxing cortical veins. A nidus will not be identified and the adjacent pial arterial anatomy will be normal.

Similarly a useful modality to identify a dAVF with cortical venous reflux is contrast-enhanced MRI as the refluxing veins will enhance and be quite preponderant (Figs. 7 and 8). While this may not readily help to identify the shunt site, it will help to raise the index of suspicion for an underlying dAVF. Venous hypertension may manifest radiographically as flair signal (Fig. 7B).

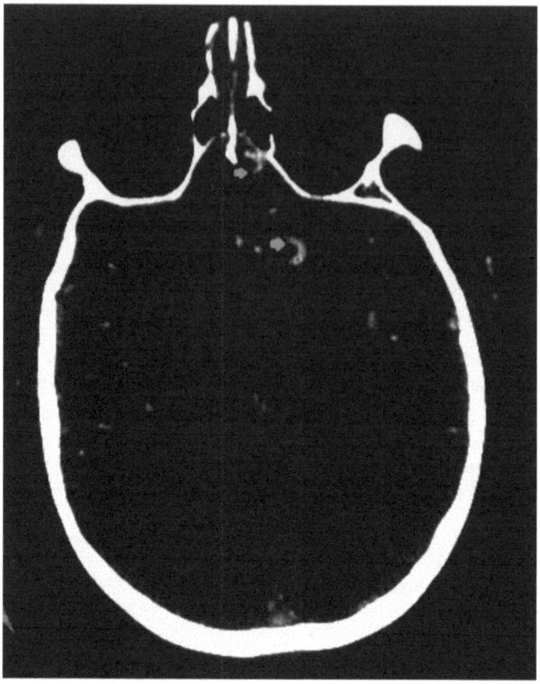

**Fig. 4** "Out of place" distended vessels along the anteroinferior frontal lobe/anterior fossa floor are suggestive of a dAVF (*arrows*).

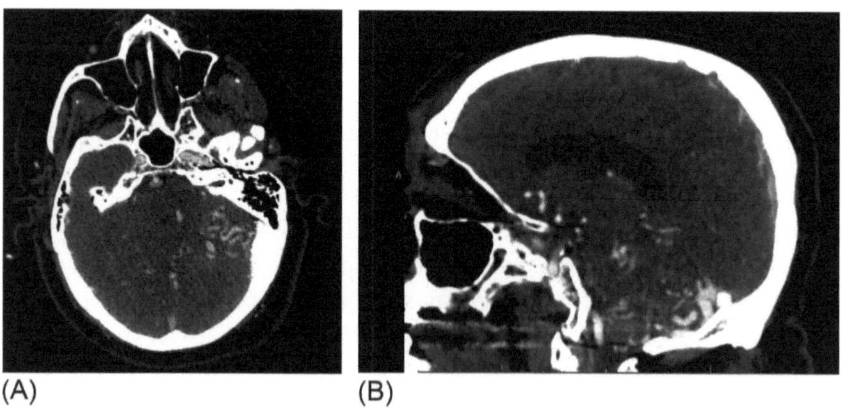

**Fig. 5** "Corkscrew veins." They are dispersed and not organized in a nidal pattern (A, axial; B, sagittal); these are the refluxing cortical veins akin to visualizing of distended veins in spinal dAVFs.

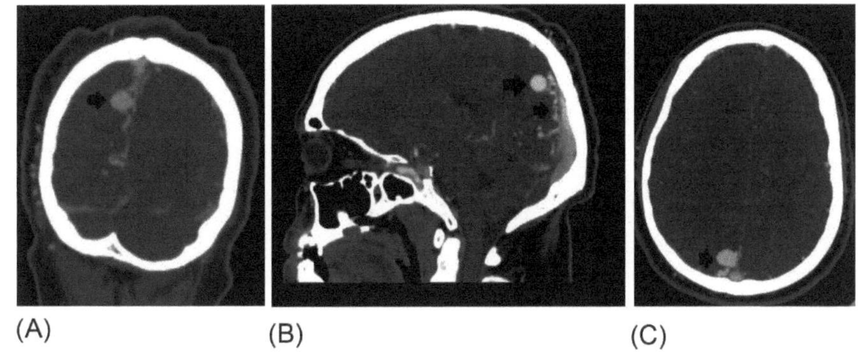

Fig. 6 Type IV superior sagittal sinus dAVF with easy identification of the varix (*arrow*; A, coronal; B, sagittal; C, axial).

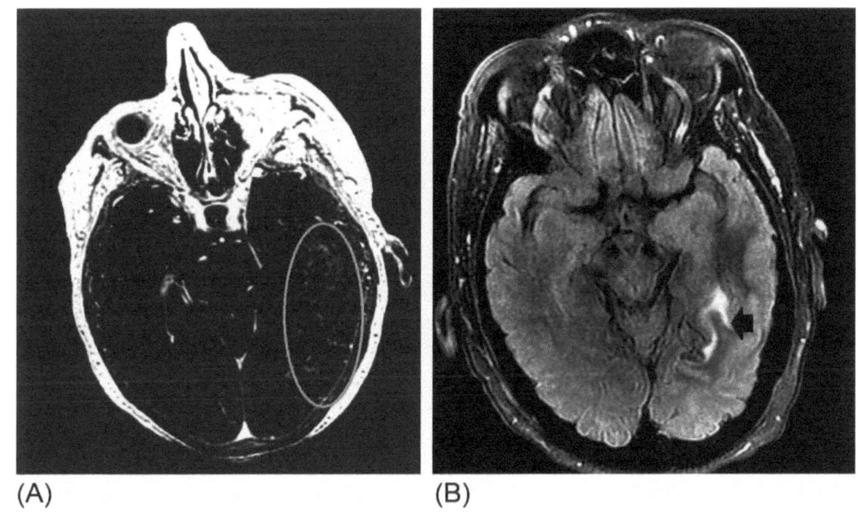

Fig. 7 Contrast enhanced T1-weighted MRI of a dAVF with cortical venous reflux (A, refluxing cortical veins) and resultant radiographic venous hypertension on flair (B, *arrow*; flair signal).

With an appropriate index of suspicion, angiographic evaluation for a dAVF mandates full six-vessel angiography. In addition, if an appropriate index of suspicion exists, and a dAVF is not identified, the practitioner should ensure there is adequate opacification of the ascending pharyngeal artery on the external carotid run; if not, a superselective run or at least a common carotid run with good visualization of the foramen magnum region should be performed (Fig. 9).

Full six-vessel angiography is mandated in the evaluation of dAVFs to identify the full gamut of feeders to the lesion. From an endovascular perspective, this allows the practitioner to evaluate for the optimal transarterial conduit.

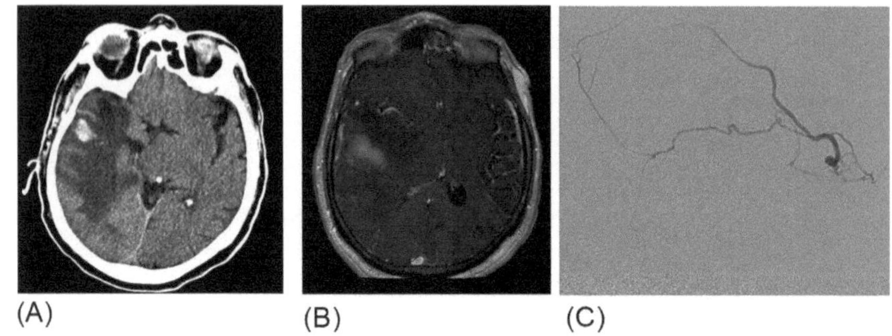

**Fig. 8** dAVF misdiagnosed first as a stroke and then a glioblastoma (A, noncontrast head CT). Contrast-enhanced T1 post-MRI demonstrates refluxing cortical veins (B) and angiography confirmed a Type III dAVF (C, lateral superselective middle meningeal run).

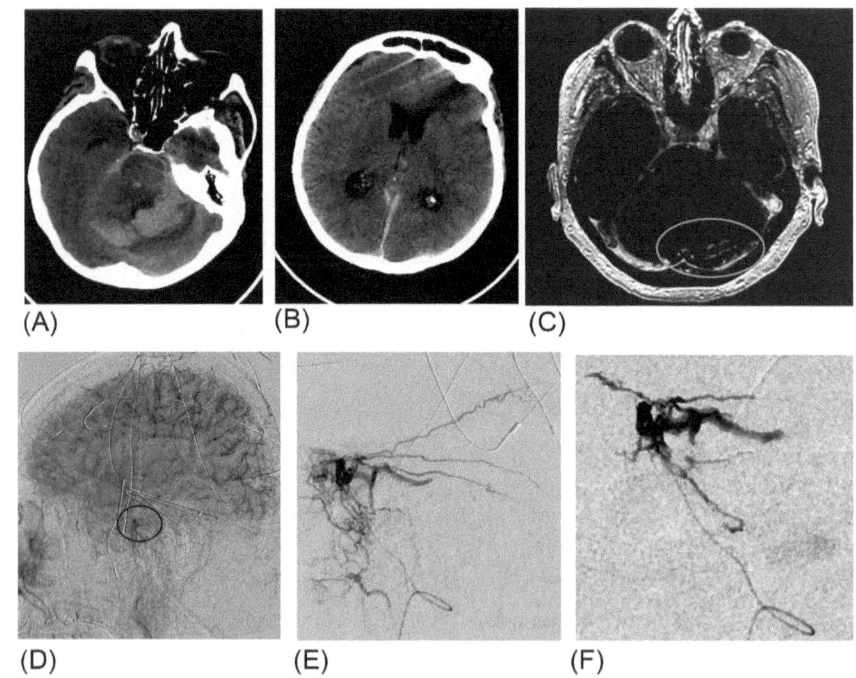

**Fig. 9** This patient presented with a spontaneous cerebellar intraparenchymal hemorrhage (A, noncontrast CT) and falcotentorial subdural hematoma (B, noncontrast CT). Contrast-enhanced MRI raised suspicion for a dAVF with enhancing serpiginous veins (C). Subtle early opacification of the posterior cavernous sinus best seen on a common carotid run (D), with superselective ascending pharyngeal angiography confirming the dAVF (E) that was treated via transarterial embolization (save fluoro of Onyx cast after first push, F).

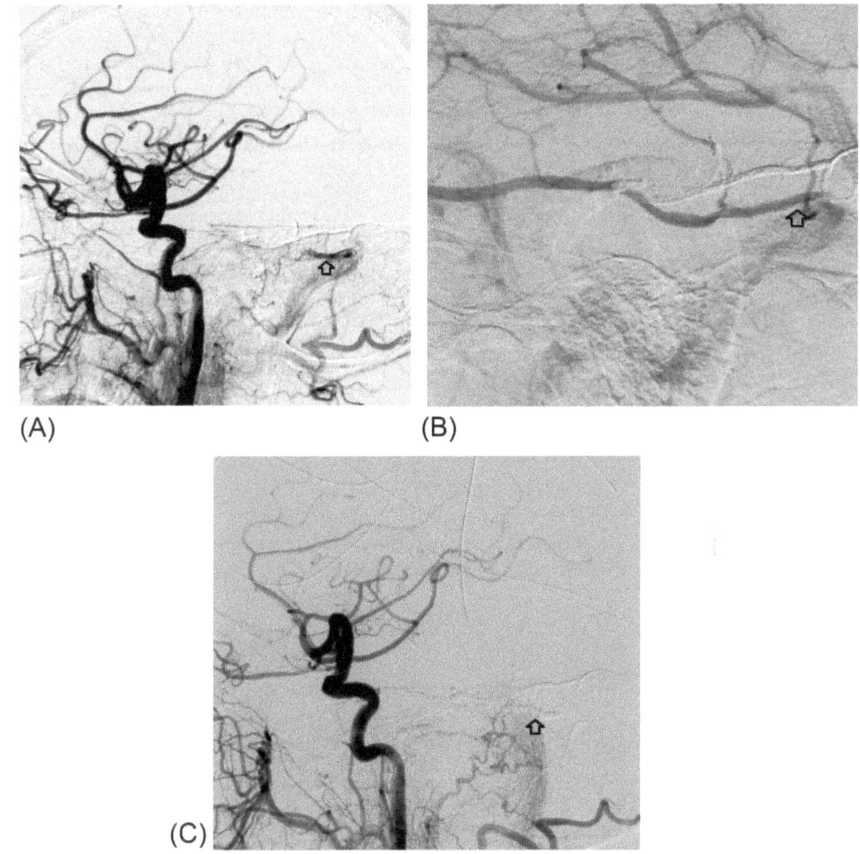

**Fig. 10** The venous outflow of this fistula (*arrow*, A) follows a similar distal course on the lateral view as this basal temporal vein in the normal venous phase (*arrow*, B). Care was taken to ensure entry into the proper vein during transvenous coil embolization (*arrow* on coil pack, C). Note can be made of an old Onyx cast as this patient had a previously embolized pial AVM and developed a de novo dAVF after craniotomy.

Generally, larger, less tortuous vessels are ideal and nontransosseous pushes are often sought (middle meningeal artery more ideal than occipital). Identification of posterior meningeal feeders and cavernous internal carotid artery (ICA) branch feeders is particularly important to understand their course and readily identify liquid embolisate refluxing towards these branches during embolization. For example, the lesion in Fig. 8 had no identifiable cavernous ICA feeders, providing a margin of safety for the transarterial approach via the clival branch of the neuromeningeal trunk. Normal venous outflow should also be evaluated to ensure there is no compromise to normal veins adjacent to the lesion, particularly germane for anticipated transvenous approaches or surgical disconnection (Fig. 10).

## References

1. Gross BA, Albuquerque FC, Moon K, McDougall CG. Evolution of treatment and a detailed analysis of occlusion, recurrence and clinical outcomes in an endovascular library of 260 dural arteriovenous fistulas. *J Neurosurg* 2017;**126**:1884–93.
2. Gross BA, Albuquerque FC, McDougall CG, et al. A multi-institutional analysis of the untreated course of cerebral dural arteriovenous fistulas. *J Neurosurg* 2018;**129**:1114–9.
3. Gross BA, Moon K, Kalani MY, et al. Clinical and anatomic insights from a series of ethmoidal dural arteriovenous fistulas at Barrow neurological institute. *World Neurosurg* 2016;**93**:94–9.
4. Pu J, Si X, Ye R, Zhang B. Straight sinus dural arteriovenous fistula presenting with reversible parkinsonism: a case report and literature review. *Medicine (Baltimore)* 2017;**96**, e9005.
5. Brito A, Tsang AC, Hilditch C, Nicholson P, Krings T, Brinjikji W. Intracranial dural arteriovenous fistula as a reversible cause of dementia: case series and literature review. *World Neurosurg* 2019;**121**:e543–53.
6. Narvid J, Do HM, Blevins NH, Fischbein NJ. CT angiography as a screening tool for dural arteriovenous fistula in patients with pulsatile tinnitus: feasibility and test characteristics. *AJNR Am J Neuroradiol* 2011;**32**:446–53.
7. Lin YH, Lin HH, Liu HM, Lee CW, Chen YF. Diagnostic performance of CT and MRI on the detection of symptomatic intracranial dural arteriovenous fistula: a meta-analysis with indirect comparison. *Neuroradiology* 2016;**58**:753–63.

# CHAPTER 4

# The natural history of cerebral dural arteriovenous fistulas

**Bradley A. Gross[a] and Rose Du[b]**
[a]Department of Neurological Surgery, University of Pittsburgh, Pittsburgh, PA, United States
[b]Department of Neurosurgery, Brigham and Women's Hospital and Harvard Medical School, Boston, MA, United States

Dural arteriovenous fistulas (dAVFs) are generally considered acquired lesions, most commonly idiopathic in nature. A known subset, however, may develop following cranial trauma, infection, venous sinus thrombosis and prior craniotomy.[1–4] The simplest approach to understanding cerebral dural arteriovenous fistula (dAVF) natural history is to understand the fistula's venous outflow. As the shunt is contained within the dural leaflets, if drainage does not connect to the subpial compartment via cortical venous reflux, the patient is not at risk of intracranial venous hypertension and/or hemorrhage. Nevertheless, these dAVFs may cause symptoms in the form of tinnitus, or, if causing ocular venous hypertension due to reflux into the superior ophthalmic vein, ocular symptoms including chemosis, proptosis, and/or frank cranial nerve palsy.

Natural history needs to account for both the natural course of the fistula itself, that is, risk of spontaneous regression versus progression, and the course of associated symptoms including tinnitus, ocular symptoms, nonhemorrhagic neurological deficits due to venous hypertension, and hemorrhage. Unfortunately, at this point, the natural course of pulsatile tinnitus is poorly understood. In some cases it can progress and become debilitating, while in a hearty minority, it can spontaneously resolve with or without concomitant spontaneous dAVF thrombosis. Current analyses are significantly limited by selection bias and treatment of dAVFs with associated debilitating pulsatile tinnitus. This is even more so true for the natural history of ocular symptoms secondary to fistulas with reflux into the superior ophthalmic vein that are nearly ubiquitously treated.

Houser et al. were the first to embolden the importance of venous drainage pattern on natural history risk in their 1972 publication.[1] They demonstrated that dAVFs with cortical venous drainage pose a risk of "aggressive symptoms" while those without were largely associated with tinnitus or nonaggressive symptoms. This was later more rigorously parsed out

*Cerebral Dural Arteriovenous Fistulas*
https://doi.org/10.1016/B978-0-12-819525-3.00003-4

in Djindjian's 1978 classification scheme[2] and the Cognard-Merland further update in 1995 (Table 1).[3] The Borden-Shucart classification eliminated the Djindjian Type IV designation in an attempt to extrapolate the scheme to spinal dAVFs.[5] While this extrapolation did not garner a following, many now refer to Djindjian Type I–III dAVFs as "Borden" Types I–III.

Most early natural history studies evaluated fistula presentation as a surrogate for natural history. They were clearly able to lineate the importance of venous drainage on aggressive symptoms and rupture rate on presentation.[3, 4] A review from 1990 that compiled 360 dAVFs from the literature combined with 17 cases of the authors' own concluded that leptomeningeal venous drainage, variceal venous dilatations, and galenic drainage were associated with an aggressive neurological presentation.[4] The Cognard classification paper noted that of 84 Type I fistulas, 83 presented with nonaggressive symptoms and none with hemorrhage. The majority of Type III fistulas presented with aggressive symptoms and Type IV fistulas had the highest hemorrhage rate on presentation with only one presenting with nonaggressive symptoms.[3]

In 2002, the University of Toronto group published two papers that evaluated dAVF history over a follow-up interval to eliminate potential presentation bias and lineate an initial estimate of annualized event rates. One paper was dedicated to dAVFs without cortical venous reflux and reported one hemorrhage over a mean follow-up of 27.9 months for 117 patients with dAVFs initially without cortical venous reflux.[6] It was noted that this patient's dAVF had converted to develop cortical venous drainage. Of fifty patients with follow-up angiography in the study, venous drainage pattern change was noted in five, with two demonstrating conversion to developing cortical venous drainage. Their other paper evaluated 20 dAVFs with cortical venous drainage, including six that were partially treated and found over

**Table 1** The Djindjian and subsequent modification, the Cognard classification.

| Djindjian Type | Cognard Type | Definition |
|---|---|---|
| I | I | Drainage into a dural venous sinus with antegrade sinus flow |
| I | IIa | Drainage into a dural venous sinus with retrograde sinus flow |
| II | IIb | Drainage into a dural venous sinus with antegrade sinus flow and cortical venous reflux |
| II | IIa + b | Drainage into a dural venous sinus with retrograde sinus flow and cortical venous reflux |
| III | III | Direct CVD |
| IV | IV | Direct CVD with vein ectasia |
| | V | Spinal perimedullary venous drainage |

a mean follow-up period of 4.3 years, a 6.9% annualized nonhemorrhagic neurologic deficit rate and 8.1% annual hemorrhage rate.[7] Annualized mortality was 10.4%.

The term "nonhemorrhagic neurologic deficit" (NHND) as it applies to cerebral dAVFs refers to the development of focal or global neurological symptoms attributable to venous hypertension. Only fistulas with cortical venous drainage may cause NHND, which may include focal neurologic deficits in the form of paresis, seizures, or masquerading symptoms such as parkinsonism or dementia.[7] Through ensuing literature, the term "symptomatic" dAVF on presentation refers to those presenting with NHND or hemorrhage. This term excludes dAVFs presenting with symptoms such as tinnitus or ocular symptoms and essentially implies that a symptomatic fistula is symptomatic due to venous hypertension.

## Presentation modality

Duffau et al. demonstrated that hemorrhagic dAVFs have a potentially high rate of early rebleeding, reporting that among 20 patients presenting with hemorrhagic dAVF, the rate of rebleeding was 35% in the first 2 weeks.[8] By definition, these are dAVFs with cortical venous drainage. This was more formally emboldened by Soderman et al.'s study that evaluated 85 patients with dAVFs with CVD, finding an annual hemorrhage rate of 1.5% for those not presenting with hemorrhage and 7.4% for those presenting with hemorrhage.[9]

The principle that a hemorrhagic vascular lesion has a higher rate of rebleeding is ubiquitous in cerebrovascular neurosurgery for aneurysms, arteriovenous malformations, and cavernous malformations as well.[10–12] Importantly the principle of a variable rebleeding rate that is highest early is suggested for dAVFs by Duffau's study and has been better demonstrated for AVMs and particularly cavernous malformations.[10–12] Should this likely principle be the case for hemorrhagic dAVFs, the few natural history studies that exist assume a constant annualized rebleed rate based on relatively short follow-up and will thus overestimate the rate of rebleeding if annualized.

Strom et al. dichotomized dAVFs with cortical venous reflux into "asymptomatic," that is, not presenting with NHND or hemorrhage and symptomatic.[13] This dichotomy-treated NHND as a relevant neurological event. It demonstrated that the combined annualized NHND plus hemorrhage rate for "asymptomatic" dAVFs with cortical venous drainage was 1.4%, increasing to 19% for symptomatic lesions. This study was limited however by short follow-up, small sample sizes (17 asymptomatic and 11 symptomatic dAVFs) and the inclusion of several partially treated lesions.[13]

## Venous ectasia

While studies were focusing on venous outflow pattern and clinical presentation, Bulters and colleagues re-invigorated the relevance of the Djindjian Type IV dAVF distinction, that is, the presence of venous ectasia in addition to direct cortical venous drainage.[14] In a study of 75 dAVFs with cortical venous drainage, they found that those with venous ectasia had a sevenfold increase in the incidence of hemorrhage (3.5% vs 27%). Venous ectasia was defined as venous dilatation of at least 5 mm and three times the diameter of the draining vein. An ensuing metaanalysis would embolden this finding.[15]

## Pooled results

Table 2 summarizes results from a pooled study of single center data and a metaanalysis (total $n = 395$)[15] along with a multicenter analysis ($n = 295$)[16] to provide annualized NHND and hemorrhage rates for dAVFs. The metaanalysis does include partially treated lesions, while the multicenter analysis only includes untreated dAVFs. Both studies reinforce the significance of venous ectasia and a four-type classification system consistent with the original Djindjian system (Fig. 1). Neither study demonstrated a significant impact of patient age or sex on event rates.

Type I dAVFs drain into a venous sinus without cortical venous drainage (Fig. 1A). The original Djindjian scheme would include Cognard Type IIa in this group as well (retrograde flow in the sinus without cortical reflux). Consistent across both studies, approximately one-half of Type I dAVFs were transverse-sigmoid and approximately one-half of transverse-sigmoid dAVFs were Type I. The second most common dAVF location for Type I lesions was cavernous (approximately one-third of cases), and marginal sinus dAVFs comprised approximately one-sixth of these cases. Approximately two-third of cavernous dAVFs collectively are Type I while approximately 80% of marginal sinus dAVFs are Type I.[16] A total of 409-lesion years of follow-up were totaled in the pooled metaanalysis and 177 in the multicenter analysis for Type I dAVFs, over which no NHND or hemorrhage events occurred if the dAVF did not develop cortical venous drainage. The rate of CVD development was 1.4% in the metaanalysis. The rate of spontaneous thrombosis was 13% in the review and 3% in the multicenter series. In the authors' anecdotal experience, acutely discovered traumatic Type I dAVFs have a considerable, though as yet unquantified, rate of spontaneous thrombosis.

**Table 2** Event rates from a pooled metaanalysis[15] and multiinstitutional analysis.[16]

| | Pooled analysis (%) | Multiinstitutional analysis (%) |
|---|---|---|
| **Type I** | | |
| Annual NHND rate | 0 | 0 |
| Annual bleed rate | 0 | 0 |
| **Type II** | | |
| Annual NHND rate | | 5 |
| Annual bleed rate | 6 | 3 |
| **Type III** | | |
| Annual NHND rate | | 6 |
| Annual bleed rate | 10 | 4 |
| **Type IV** | | |
| Annual NHND rate | | 5 |
| Annual bleed rate | 21 | 9 |
| **Type II–IV asymptomatic presentation** | | |
| Annual NHND rate | 0 | 2 |
| Annual bleed rate | 2 | 3 |
| **Type II–IV NHND presentation** | | |
| Annual NHND rate | 20 | 23 |
| Annual bleed rate | 10 | 3 |
| **Type II–IV hemorrhagic presentation** | | |
| Annual NHND rate | 0 | 0 |
| Annual rebleed rate | 46 | 46 |

Type II dAVFs drain into a venous sinus with cortical venous reflux (Fig. 1B). Like Type I dAVFs, approximately one-half of Type II dAVFs are transverse-sigmoid and approximately one-half of transverse-sigmoid dAVFs are Type II. The second most common location for Type II dAVFs is cavernous. Type III dAVFs drain directly into a cortical vein (Fig. 1C). Common locations include tentorial, petrosal, superior sagittal sinus/convexity and anterior fossa. Type IV dAVFs drain directly into a cortical vein with venous ectasia (Fig. 1D) with similar typical locations as Type III dAVFs. However, venous ectasia is proportionately most common among anterior fossa dAVFs. Interestingly, a male sex predilection was noted among Type III and IV dAVFs in both analyses[15, 16]; this is akin to spinal

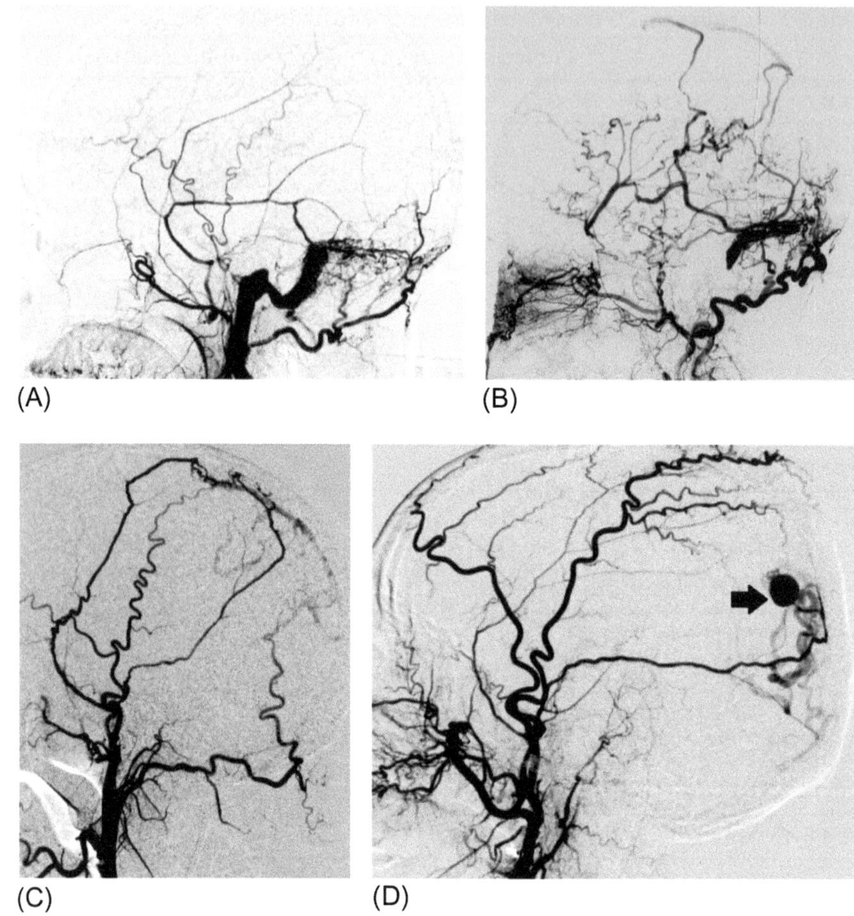


**Fig. 1** Djindjian Type I (A), Type II (B), Type III (C), and Type IV (D) dAVFs.

dAVFs.[17–19] In the pooled analysis, Type IV dAVFs harbored a 21% annual rate of hemorrhage; this was 9% in the multicenter study. It is sensible that Types II–IV dAVFs in the multicenter study harbored similar annual NHND rates (5%–6%) as the mere presence of cortical venous reflux is the etiologic cause of venous hypertension with less influence from venous morphology/dilation.

As both studies reinforce, "symptomatic" versus "asymptomatic" presentation does not entirely explain subsequent event rates. Patients with Type II–IV dAVFs that do not present with NHND or hemorrhage do have the lowest subsequent annual event rates (0%–2% annual NHND rate and 2%–3% annual hemorrhage rate). It is important however to stratify presentation by NHND versus hemorrhage. Patients presenting with NHND had the highest subsequent annualized NHND rates (20%–23%) but relatively lower annualized hemorrhage rates (3%–10%). Patients presenting with hemorrhage were most likely to rebleed but had 0% annualized NHND rates. As aforementioned, the rebleed rate (annualized to 46%) may be an overestimate as a result of early hemorrhage clustering that is annualized.

## Concluding remarks

Patients with dAVFs without cortical venous drainage (Type I) may present with tinnitus or ocular symptoms secondary to ocular venous hypertension. Unless these dAVFs develop cortical venous drainage (1.4% risk), they do not pose a risk of NHND or hemorrhage. Among dAVFs with cortical venous drainage (Types II–IV), venous ectasia (Type IV) is likely a significant risk factor for hemorrhage. In addition, NHND presentation is a significant risk factor for subsequent NHND, and hemorrhagic presentation is a significant risk factor for subsequent hemorrhage. While age and sex likely do not impact dAVF natural history, a prospective analysis of the impact of antiplatelet or anticoagulant medication may prove quite interesting given the balance of a potential beneficial effect on venous hypertension weighed against the intuitive adverse impact of blood thinning. Other interesting factors to evaluate are the impact of pregnancy (though mean age of patients with dAVFs is 50–60 years), typical medical comorbidities and smoking status.

## References


1. Houser OW, Baker HL, Rhoton AL, Okazaki H. Intracranial dural arteriovenous malformations. *Radiology* 1972;**105**:55–64.
2. Djindjian R, Merland JJ. *Superselective angiography of the external carotid artery.* Berlin/Heidelberg/New York: Springer-Verlag; 1978.
3. Cognard C, Gobin YP, Pierot L, et al. Cerebral dural arteriovenous fistulas: clinical and angiographic correlation with a revised classification of venous drainage. *Radiology* 1995;**194**:671–80.
4. Awad IA, Little JR, Akarawi WP, Ahl J. Intracranial dural arteriovenous malformations: factors predisposing to an aggressive neurological course. *J Neurosurg* 1990;**72**:839–50.
5. Borden JA, Wu JK, Shucart WA. A proposed classification scheme for spinal and cranial dural arteriovenous fistulous malformations and implications for treatment. *J Neurosurg* 1995;**82**:166–79.
6. Satomi J, van Dijk JM, Terbrugge KG, Willinsky RA, Wallace MC. Benign cranial dural arteriovenous fistulas: outcome of conservative management based on the natural history of the lesion. *J Neurosurg* 2002;**97**:767–70.
7. van Dijk JM, ter Brugge KG, Willinsky RA, Wallace MC. Clinical course of cranial dural arteriovenous fistulas with long-term persistent cortical venous reflux. *Stroke* 2002;**33**:1233–6.

8. Duffau H, Lopes M, Janosevic V, et al. Early rebleeding from intracranial dural arteriovenous fistulas: report of 20 cases and review of the literature. *J Neurosurg* 1999;**90**:78–84.
9. Soderman M, Pavic L, Edner G, Holmin S, Andersson T. Natural history of dural arteriovenous shunts. *Stroke* 2008;**39**:1735–9.
10. Hernesniemi JA, Dashti R, Juvela S, Vaart K, Niemela M, Laakso A. Natural history of brain arteriovenous malformations: a long-term follow-up study of risk of hemorrhage in 238 patients. *Neurosurgery* 2008;**63**:823–31.
11. Gross BA, Du R. Rate of rebleeding of arteriovenous malformations in the first year after rupture. *J Clin Neurosci* 2012;**19**:1087–8.
12. Barker II FG, Amin-Hanjani S, Butler WE, et al. Temporal clustering of hemorrhages from untreated cavernous malformations of the central nervous system. *Neurosurgery* 2001;**49**:15–24.
13. Strom RG, Botros JA, Refai D, et al. Cranial dural arteriovenous fistulae: asymptomatic cortical venous drainage portends less aggressive clinical course. *Neurosurgery* 2009;**64**:241–8.
14. Bulters DO, Mathad N, Culliford D, Millar J, Sparrow OC. The natural history of cranial dural arteriovenous fistulae with cortical venous reflux—the significance of venous ectasia. *Neurosurgery* 2012;**70**:312–9.
15. Gross BA, Du R. The natural history of cerebral dural arteriovenous fistulae. *Neurosurgery* 2012;**71**:594–603.
16. Gross BA, Albuquerque FC, McDougall CG, et al. A multi-institutional analysis of the untreated course of cerebral dural arteriovenous fistulas. *J Neurosurg* 2018;**129**:1114–9.
17. Ropper AE, Gross BA, Du R. Surgical treatment of Type I spinal dural arteriovenous fistulas. *Neurosurg Focus* 2012;**32**, E3.
18. Steinmetz MP, Chow MM, Krishnaney AA, et al. Outcome after the treatment of spinal dural arteriovenous fistulae: a contemporary single-institution series and meta-analysis. *Neurosurgery* 2004;**55**:77–87. discussion 87–78.
19. Saladino A, Atkinson JL, Rabinstein AA, et al. Surgical treatment of spinal dural arteriovenous fistulae: a consecutive series of 154 patients. *Neurosurgery* 2010;**67**:1350–7. discussion 1357–1358.

# CHAPTER 5

# Transarterial embolization

**David I. Bass[a], Christopher C. Young[a], Rajeev D. Sen[a], and Michael R. Levitt[a,b,c,d]**

[a]Department of Neurological Surgery, University of Washington, Seattle, WA, United States
[b]Department of Radiology, University of Washington, Seattle, WA, United States
[c]Department of Mechanical Engineering, University of Washington, Seattle, WA, United States
[d]Stroke & Applied Neuroscience Center, University of Washington, Seattle, WA, United States

The primary goal of treatment of dural arteriovenous fistulas (dAVF) is to occlude the fistulous connection between the arterial and venous systems. More proximal arterial ligation may allow collateral feeders to form, while more distal venous ligation may exacerbate venous hypertension, thereby provoking a venous infarct or hemorrhage.[1] For most dAVFs that require treatment, the initial step in management is often endovascular embolization,[2, 3] but the American Heart Association has not made a formal recommendation regarding treatment modality.[4] Early endovascular strategies preferred transvenous embolization; however, advancements in technology, including the development of liquid embolic agents such as *n*-butyl cyanoacrylate (nBCA) and ethylene-vinyl alcohol copolymer (EVOH), have provided new tools for transarterial embolization.

Onyx (Medtronic, Minneapolis, MN), EVOH dissolved in dimethylsulfoxide (DMSO), has emerged as the embolic agent of choice for the treatment of dAVFs[5, 6] in large part due to its cohesive nature in comparison with other adhesive liquid embolic agents.[7] Onyx was experimentally introduced in swine models for endovascular embolization in the 1990s[8] and was approved by the Food and Drug Administration in 2005 for preoperative embolization of cerebral arteriovenous malformations. Subsequently the application of Onyx expanded to the off-label treatment of dAVFs,[9] fundamentally redefining endovascular management of these lesions.

This chapter discusses the technical details of the transarterial approach for embolization of cranial dAVFs with Onyx. We begin with patient selection and material properties, discuss treatment outcomes and complications associated with this strategy, and then conclude with several illustrative cases to review key learning points.

*Cerebral Dural Arteriovenous Fistulas*
https://doi.org/10.1016/B978-0-12-819525-3.00008-3

## Patient selection for surgical approach

Patients with high-risk dAVFs, as defined by cortical venous reflux (e.g., Type II–III),[10] require treatment in most cases to prevent neurological deficit or hemorrhage,[4] especially in the setting of cortical venous hypertension. Patients with "benign" symptoms such as pulsatile tinnitus or ophthalmological involvement may also require treatment for symptom resolution. A transvenous approach may be indicated, either alone or in combination with transarterial embolization when the lesion is formed by multiple small feeding arteries requiring separate catheterizations to successfully occlude the fistula,[11] or in cases when normal venous drainage is distinct from that of the dAVF, such as a dAVF draining into an isolated transverse sinus. Tortuous vascular anatomy within the neck can elevate the risk of endovascular intervention, and lack of vascular access can altogether preclude angiography. An open surgical approach may be more appropriate for such patients, and for patients that have lesions with prominent osseous arterial feeders,[2] or dAVF arising from arteries supplying other critical structures that may be put at risk from embolization, such as ethmoidal dAVF arising from branches of the ophthalmic artery.[12] Endovascular and open approaches can be combined in exceptional circumstances during a single session to directly access an otherwise inaccessible arterial feeder via a craniotomy.[13]

A transvenous approach was initially the preferred endovascular route of access because it had higher rates of achieving a complete occlusion with available embolic agents such as coils.[14] Part of the success of the transvenous approach stems from the ability to sacrifice the sinus, but this strategy may increase the risk of embolization or venous infarction if the fistulous outflow is shared with normal cortical venous drainage.[3, 15] In comparison, proximal transarterial approaches had poor long-term durability, presumably due to recanalization of the fistula by arterial collaterals that developed following proximal artery embolization. Therefore the transarterial approach had been reserved for use in combination with other strategies,[14] such as preoperative embolization to help limit blood loss during open ligation,[2] or in complement with a transvenous approach to improve rates of complete endovascular obliteration.[5]

With the introduction of nBCA, treatment shifted toward a transarterial approach for embolization. When compared with proximal embolization, liquid embolics improved distal embolization into the fistula itself and increased the permanence of the occlusion.[14] The subsequent development of Onyx continued to improve cure rates via a transarterial approach in a single session without additional venous catheterization,[16] and as a result, transarterial embolization of dAVFs has emerged as the preferred strategy over the past two decades.[5, 17] A transarterial approach works particularly well for lesions with direct meningeal arterial feeders[18] and for those with direct cortical venous reflux.[19] In addition, some groups report good results in dAVFs with indirect cortical venous reflux.[17] Furthermore, good safety and efficacy have been demonstrated with transarterial embolization with Onyx for lesions with indirect cortical venous drainage when a sinus occlusion prevents venous access, and in skull base lesions previously thought too dangerous for endovascular therapy due to risk of reflux.[19, 20] However, transarterial embolization should be avoided if the feeding vessel arises from or collateralizes with the intracranial circulation, or if embolization may compromise the arterial supply to a cranial nerve.[5]

## Material properties and selection

Besides embolic aneurysm coils, three injectable embolization products are most often used in the transarterial treatment of dAVFs: Onyx, nBCA, and PVA. Onyx is available in two concentrations of EVOH, Onyx 18 (6% EVOH) with a nominal viscosity of 18 cST and Onyx 34 (8% EVOH) with a nominal viscosity of 33 cST.[21] The lower viscosity formulation allows for deeper penetration of smaller feeders while the higher viscosity formulation is intended for high-flow environments. However, selection criteria for either formulation have not been formally established.[7] The manufacturer recommends augmenting Onyx embolization with detachable coils when embolizing high-flow fistulae (more than 200 mL/min), or in arterial feeding pedicles more than 3 mm in diameter.[21]

EVOH must be delivered in an organic solvent, such as DMSO, to prevent premature copolymerization.[21] As a result, additional precautions must be taken to prevent angiotoxicity from chemical exposure, such as slow, controlled delivery (0.3 mL over 40 s).[8, 22] If vasospasm occurs, it is unlikely to be clinically significant, and typically lasts only 10–20 min.[22] In contrast, a larger bolus of DMSO delivered through a quick injection (1.0 mL in less than 15 s) causes endoluminal granulomatous inflammation, severe vasospasm, thrombotic occlusions, and transmural necrosis.[22] Hypotension, abnormal heart rhythms, and death have been reported in extreme circumstances.

During injection, DMSO quickly dissipates allowing the precipitate to form a "skin" at the interface between the copolymer and blood, or any other aqueous solution.[8,21] The core of the injection remains liquid for up to 5 min before it solidifies.[21] This phenomenon results in the "lava-like" flow pattern as described by many groups, allowing for a longer injection period that can be stopped and restarted within minutes.[15,16,23] Final solidification forms a soft, spongy plug that fills the entire lumen of the blood vessel, quickly achieving complete occlusion without breaking into smaller embolic fragments.[24] As the copolymer is cohesive rather than adhesive, it does not stick to the blood vessel walls.[8,25] This characteristic is thought to decrease thrombogenicity and local inflammation in comparison with embolic glue.[24] Further the cohesive nature of Onyx is thought to yield more predictable outcomes in comparison with other embolic agents that have distinctly different behaviors with relatively more stochastic results even after extensive training.

nBCA is an embolic glue, mixed with ethiodol, that polymerizes upon contact with aqueous solutions.[8] It quickly forms a hard cast which then shrinks away from the wall of the vessel.[24] Its rapid polymerization increases the risk of premature occlusion of the catheter or proximal vessel, preventing good distal penetration into the fistula.[26] When lesions are embolized preoperatively, surgeons have found that vessels occluded with nBCA are more brittle and harder to coagulate than those embolized with Onyx.[24] This finding may reflect protein denaturation within the vessel walls caused by heat produced from nBCA polymerization.[8] Finally, rapid polymerization creates a risk of leaving a retained catheter due to adhesion to the vessel wall.

One potential advantage of nBCA over Onyx is that nBCA's viscosity can be titrated by the operator by increasing the concentration of ethiodol in the mixture.[27] Appropriate titration of viscosity is a critical step in successful embolization with this polymer, but it requires extensive experience to achieve optimal results.

PVA consists of spongy microparticles available in different sizes. The particles form an occlusive mass through physical aggregation[28] or the forming of reversible bonds, rather than forming an adhesive or cohesive glue. While there is no risk of retained catheters, PVA suffers from two disadvantages compared with other embolic agents: lack of durability and possible traversal of the lesion into the venous system, with a corresponding risk of pulmonary embolism.[26] PVA embolization may also be associated with increased risk of perioperative hemorrhage due to venous outflow obstruction, and it is more likely to require augmentation with coils to achieve a similar degree of success when embolizing vascular lesions.[26]

## Procedure in detail

Onyx embolization of dAVF should be performed under general anesthesia with bi-plane fluoroscopy to achieve optimal results, defined as complete occlusion of the fistula without off-target embolization. Patients should be fully heparinized with an activated clotting time of at least 250 s. Tantalum powder must be suspended in the Onyx solution for adequate visualization during embolization. To do so, the manufacture recommends vigorously shaking the vial for at least 20 min (usually with the aid of a commercially available vial agitator) until just prior to injection. A delay in injection may impair visualization as the tantalum powder can settle out of solution.

A DMSO-compatible microcatheter should be flushed with saline and then the dead space filled with DMSO. The hub of the catheter should be washed with DMSO to prevent premature solidification of Onyx. Next, a syringe filled with Onyx is attached to the catheter in a manner that prevents air bubbles from forming. A common strategy is to hold the catheter vertically, overfill the luer hub with DMSO, and then connect the syringe. The syringe is pointed vertically to create a clear interface between the Onyx and the DMSO.

The critical step in successful Onyx embolization is wedging of the microcatheter within the distalmost aspect of the arterial feeder, as close to the fistula as possible.[29] Once the microcatheter is positioned, embolization can precede using the "plug and push" technique.[7] Thumb pressure should be applied to the end of the syringe, injecting at a rate of 0.16 mL/min, and up to 0.30 mL/min. The goal is to prevent angiotoxicity and vasospasm by controlling the rate the blood vessel wall is exposed to DMSO as the DMSO clears from the catheter. Once Onyx reaches the blood stream, a brief, 1–2 min pause allow for the formation of a plug, promoting a more predictable embolization pattern.[7,16] A pause longer than 2 min risks occlusion of the microcatheter, at which point, the pressure required to overcome the plug may rupture the microcatheter.

Embolization should proceed slowly until venous outflow is occluded, and then another pause should be taken.[29] Additionally, 1–2 min pause should be taken at any point to prevent off-target embolization if reflux along the catheter is observed while working in a branch off an artery in the intracranial circulation, such as the internal carotid artery or basilar artery, or if Onyx is seen to flow into normal venous architecture.[16,25] Limiting the degree of reflux also helps to minimize the risk of catheter retention.[27] Once the venous outflow has been occluded, resuming the injection results in preferential backfilling of feeding arteries if necessary, ultimately achieving complete obliteration of the dAVF.[29] It is essential for the embolization material to completely penetrate and fill the venous pouch of the dAVF. If this is not achieved, for instance, through embolization of only the proximal arterial feeders, then the fistula is likely to persist or recur.

At the end of the procedure, a brief pause followed by small aspiration separates the Onyx within the microcatheter from the cast within the blood vessels. If a catheter is difficult to withdraw, it is most likely related to excessive reflux, vasospasm, or tortuous arterial anatomy.[21] One strategy for separating the catheter is to apply 3–4 cm of traction along the catheter for several seconds, then release and repeat as appropriate. Throughout microcatheter removal, the saline flush of the guide catheter should be halted and continuous suction should be applied on the Onyx syringe attached to the microcatheter, to prevent fragments of solidified Onyx from breaking off the microcatheter and being propelled by the flush of the guide catheter back into the circulation.

## Outcomes

The largest series to date of transarterial dAVF embolization was published by Gross et al.[30] who reported on 260 dural AVFs treated initially with endovascular embolization, 173 treated with Onyx and 61 treated with a transarterial approach only. Cure rates with Onyx via a transarterial approach were reported as 74%, with cure defined as persistent occlusion at a median follow-up of 4 months. Other related studies report similar angiographic outcomes with cure rates ranging between 50% and 90%.[5, 16, 19, 31–33] One study compared clinical and radiographic outcomes of intentional sinus occlusion versus sinus preservation.[34] These authors found that cure rates were higher for the sinus occlusion group, however, this was accompanied by a significantly higher rate of complication. Despite the increase in complication rates, there was no significant difference in long-term patient-reported quality of life outcomes between the two groups.

Research describing recurrence rates after transarterial embolization of dural AVFs is sparse. The few studies that evaluate long-term outcomes are heterogeneous in treatment modality, including transvenous or combined approaches, stereotactic radiosurgery, and microsurgery. Ambekar et al.[35] found that the odds of recurrence dropped significantly from 60% at 6 months to 10% at 24 months. Chandra et al.[6] confirmed these results, with all recurrences in their cohort occurring within 6 months of initial treatment. As with recurrence rates, clinical outcomes are inconsistently reported. The most common presenting symptoms for dural AVFs are pulsatile tinnitus or ocular symptoms such as proptosis, and headaches. The literature reports a wide range of rates of clinical improvement, ranging from 11% to 95%. In these studies, however, symptom worsening after angiographic cure is rare.[5, 34–36] Ng et al.[36] found that 95% of patients in their study had improvement in their symptoms after treatment; however, their sample size was small.

## Complications

The major complications following transarterial Onyx embolization of dural AVFs are inadvertent migration of embolisate, microcatheter retention, postprocedural venous hemorrhagic stroke, and complications associated with all endovascular procedures including vessel injury, cerebral infarct, or access site complications such as retroperitoneal hemorrhage.

Overall reported rates of complication vary, ranging from 0% to 18%.[5, 6, 18, 33, 37, 38] The most common cause of permanent neurologic deficit reported is due to embolization of microvasculature to the facial nerve resulting in partial facial paralysis. The other cranial neuropathy reported after transarterial embolization is neuropathy of the trigeminal nerve causing hemifacial anesthesia. Severe complications such as cranial neuropathies, ischemic stroke, and venous hemorrhagic strokes are reportedly rare, occurring in less than 2% of cases. Interestingly, one study found increased complication rates of 33% in cases of intentional sinus occlusion, pointing toward preservation of venous anatomy when possible.[34] Catheter-related complications rates were as high as 12% but all were asymptomatic.[5, 39]

Injection of Onyx into the normal cerebral venous system or reflux into normal intracranial arteries can result in devastating consequences including hemorrhage or ischemic stroke. Immediate management begins with cessation of the injection and re-evaluation of vascular anatomy with a repeat angiographic run. Onyx can be removed from a vessel using suction aspiration or stent retriever devices, in a manner similar to mechanical thrombectomy for acute ischemic stroke. Generally, proceeding with embolization is acceptable after addressing such a complication, as repeated inadvertent injection is unlikely. However, it is critical to understand important extracranial-intracranial anastomoses and the microvascular supply to the cranial nerves to minimize the risk of injury.[40] Two vessels to highlight are the ethmoidal artery, which anastomoses with the ophthalmic artery, and the petrous branch of the middle meningeal artery, which supplies the trigeminal and facial nerves. In addition to cortical and cranial nerve infarcts, injection into the venous system can rarely embolize to the pulmonary system, leading to acute hemodynamic changes. Overall management of this group of complications involves hemodynamic and neurologic supportive care.

Along the same lines, occlusion of the draining venous sinus can occur during Onyx embolization. While sometimes intentional, if done inadvertently without appreciation of the patient's venous anatomy (especially in the setting of shared venous drainage between the dAVF outflow and the patient's normal cerebral venous outflow), this can result in cortical venous hypertension and hemorrhagic stroke. This highlights the importance of initial diagnostic angiographic runs with late venous phase, well after the venous architecture of the dAVF is seen.

Management of catheter retention involves gentle traction on the catheter to disengage it from the Onyx. This must be done with extreme care so as not to disengage the Onyx itself or damage the surrounding vasculature. If the catheter cannot be removed, it is cut at the access site, and the portion within the body will eventually endothelialize. This complication has become less common with the advent and use of microcatheters with detachable tips allowing for retention of only the portion of the catheter within the Onyx. Nevertheless, proper technique with slow injection without excess reflux over the distal catheter tip should always be used to minimize risk.

## Illustrative cases

### Case 1: Unilateral transarterial middle meningeal artery embolization of tentorial Type III dAVF

A 69-year-old woman presented with 3 days of headache, nausea, and vomiting and was found to have a left side cerebellar intraparenchymal hemorrhage (IPH), crowding of the foramen magnum, and early hydrocephalus (Fig. 1A). A computed tomography angiogram (CTA) of her head demonstrated abnormal arterial vessels emanating from the basal cisterns, with a focal vascular dilation within the center of the IPH, presumed to be the source of the hemorrhage (Fig. 1B). The patient underwent an emergent decompressive suboccipital craniectomy with C1 laminectomy and insertion of an external ventricular drain.

The next day, under general anesthesia, diagnostic angiography showed patent venous sinuses (Fig. 1C) and a left tentorial Type III dAVF supplied

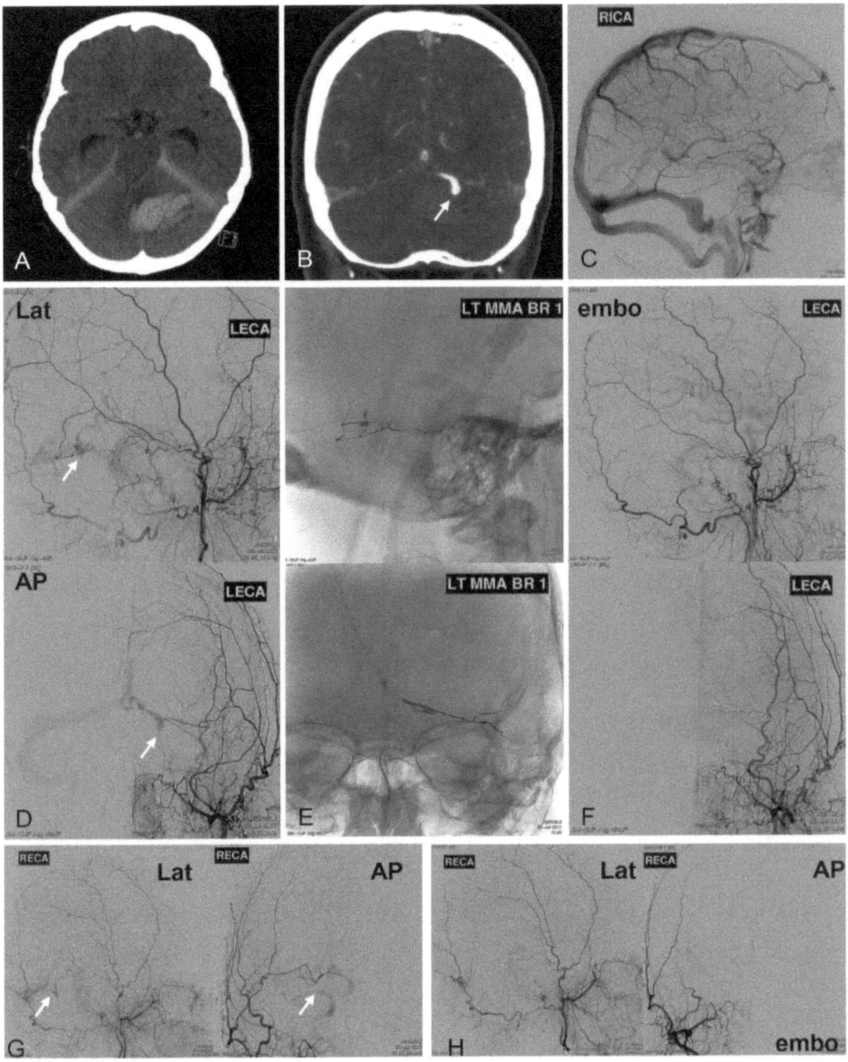


**Fig. 1** Case 1: Unilateral transarterial middle meningeal artery embolization of tentorial Type III dAVF. The patient presented with a left cerebellar intraparenchymal hemorrhage (IPH) resulting in obstructive hydrocephalus (A). A CTA demonstrated abnormal arterial vessels emanating from the basal cisterns. A focal vascular dilation within the center of the IPH was the presumed source of the hemorrhage (B, *arrow*). Diagnostic angiography showed patent venous sinuses (C) and a left tentorial Type III dAVF supplied by bilateral transosseous external carotid artery (ECA) branches (D, *arrow*); the left posterior parietal branch of the middle meningeal (MMA) and right retroauricular branch of the occipital artery. Superselective angiography of the posterior parietal branch of the left MMA demonstrated filling of the fistula with no en-passage vessels. Embolization with *(Continued)*

by bilateral external carotid artery branches (Fig. 1D): the left posterior parietal branch of the middle meningeal artery (MMA) and right retroauricular branch of the occipital artery. A 6-French (Fr) Benchmark and 5-Fr Select catheters were taken into the left external carotid artery (ECA). Under roadmap guidance, a Marathon microcatheter (Medtronic, Minneapolis, MN) and Mirage microwire (Medtronic, Minneapolis, MN) were navigated into the distal aspect of the left MMA. Superselective angiography demonstrated filling of the fistula with no en-passage vessels. At this point, the catheter was flushed with DMSO, and then Onyx 34 was injected under negative roadmap guidance to achieve penetration of the entire fistulous connection and the common venous pouch (Fig. 1E and F). Angiographic views of the right ECA, including a dedicated superselective occipital artery run, confirmed that there was no residual supply to the fistula (Fig. 1G and H).

The patient was extubated the next day. She remained in Neuro ICU until her external ventricular drain could be weaned, and she was discharged to a nursing facility 30 days after admission. One year follow-up angiogram showed durable occlusion of the dAVF without evidence of recurrence. The patient eventually made a good functional recovery without neurological deficits.

## Case 2: Bilateral transarterial middle meningeal artery embolization of convexity Type III dAVF

A 76-year-old man with severe aortic stenosis and coronary artery disease on dual antiplatelet therapy for coronary artery stents presented with a single seizure consisting of transient confusion and expressive aphasia following coronary stenting. A CTA showed prominent left parietal vessels suspicious for a dAVF (Fig. 2A). Diagnostic cerebral angiography demonstrated a left parietal convexity Type III dAVF supplied by branches of bilateral MMA and occipital arteries (Fig. 2B–D). Given the presence of high-risk features

---

**Fig. 1, cont'd** Onyx 34 achieved occlusion of the entire fistulous connection as well as penetration into the common venous pouch (E). Final angiographic views demonstrated complete obliteration of the fistula (F). Preembolization angiographic runs of the right ECA demonstrated filling from the occipital artery to the fistula (G, *arrow*). After embolization of the left MMA branch, repeat right ECA angiographic views confirmed that there was no residual supply to the fistula (G, H). *AP*, anteroposterior; *embo*, postembolization; *Lat*, lateral; *LECA*, left external carotid artery; *R MMA BR1*, right middle meningeal artery branch 1; *RECA*, right external carotid artery; *RICA*, right internal carotid artery.

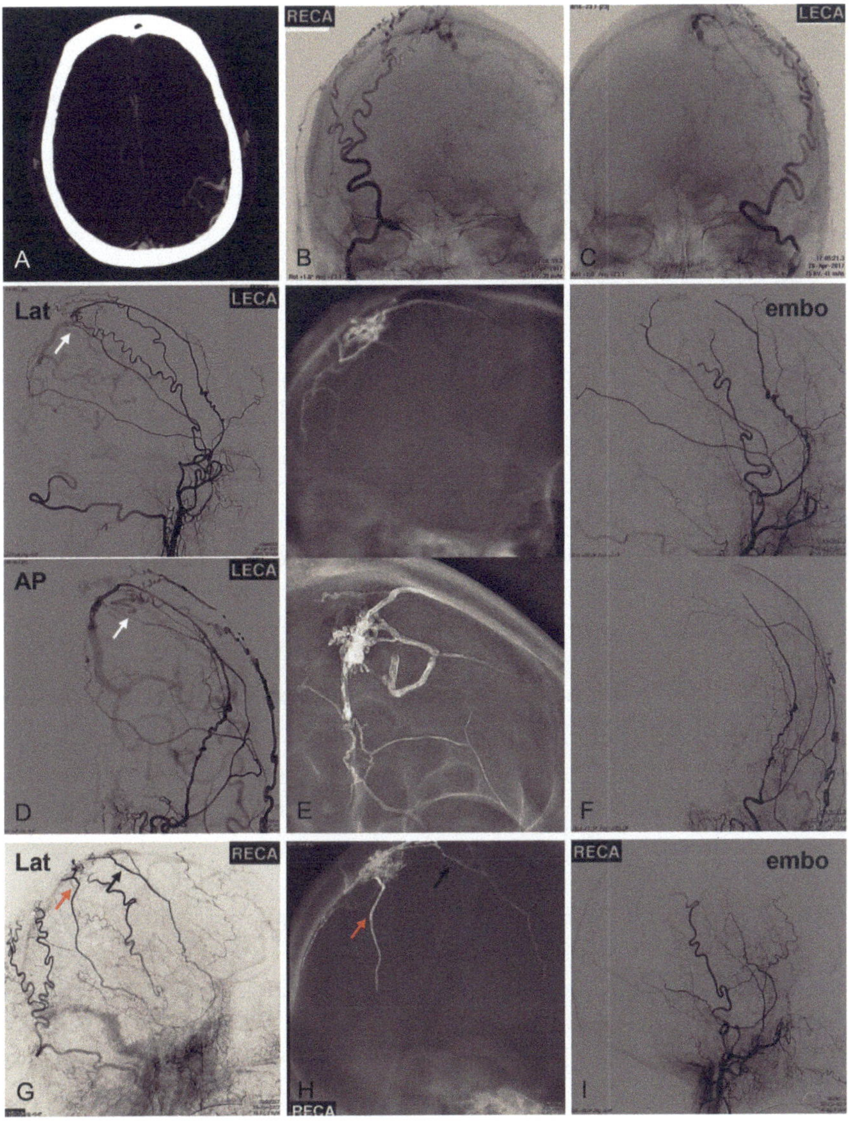


**Fig. 2** Case 2: Bilateral transarterial middle meningeal artery embolization of convexity Type III dAVF. A 76-year-old man was found to have prominent left parietal vessels on a CTA after a single seizure (A). Diagnostic cerebral angiography demonstrated a left parietal convexity Borden Type III dAVF supplied by branches of bilateral middle meningeal arteries (MMA) and occipital arteries (B–D, *arrows*). Superselective angiography demonstrated filling of the fistula through the parietal branch off the left MMA. Embolization with Onyx 34 achieved complete opacification of the dAVF (E, F). However, repeat angiographic views demonstrated persistent supply to a small portion of the fistula from

*(Continued)*

and the need for continued dual antiplatelet therapy, the decision was made to attempt Onyx embolization to occlude the dAVF.

An H1 catheter was placed into the left common carotid artery and then exchanged over a stiff glidewire for a 6-Fr Benchmark guide catheter, which was advanced into the proximal left internal maxillary artery. Under roadmap guidance, a Marathon microcatheter and Mirage microwire were navigated into the parietal branch of the left MMA. Superselective angiography demonstrated supply to the dAVF. Onyx 34 was gently injected under negative roadmap guidance until opacification of the fistula was complete (Fig. 2E and F).

Angiographic views from the right ECA demonstrated persistent supply to a small portion of the fistula from right MMA branches (Fig. 2G). Further embolization with Onyx 34 was performed via the parietal branch of the MMA (Fig. 2H, black and red arrows) and subsequent angiographic views demonstrated complete angiographic embolization of the fistula (Fig. 2I). The patient recovered well from the procedure without adverse sequelae. The 1-year follow-up angiogram showed no evidence of residual or recurrent dAVF and no further clinical events.

## Case 3: Transarterial and transvenous embolization of complex transverse sigmoid junction Type III dAVF

A 64-year-old woman presented with several months of progressive neurological decline, including frequent falls, confusion, and intermittent expressive aphasia. An MRI of the brain demonstrated serpiginous vessels, suspicious for a vascular malformation. A diagnostic cerebral angiogram performed at another institution reported a complex left transverse sigmoid junction dAVF with significant cortical venous reflux. Because of the symptomatic and high-grade nature of the lesion, the patient was referred for endovascular embolization.

Angiography showed a complex left transverse sigmoid junction dAVF with arterial supply from the right occipital artery (Fig. 3A), left posterior

---

**Fig. 2, cont'd** right MMA branches (G, *arrows*). After further embolization of the anterior branch of the right MMA (H, *black arrow*), a small portion of the fistula remained. Superselective angiography of the parietal branch off the right MMA demonstrated further supply to the dAVF which was further embolized (H, *red arrow*). Final angiographic views demonstrated complete angiographic embolization of the fistula (I). *AP*, anteroposterior; *embo*, postembolization; *Lat*, lateral; *LECA*, left external carotid artery; *LICA*, left internal carotid artery; *LVERT*, left vertebral artery; *RECA*, right external carotid artery.

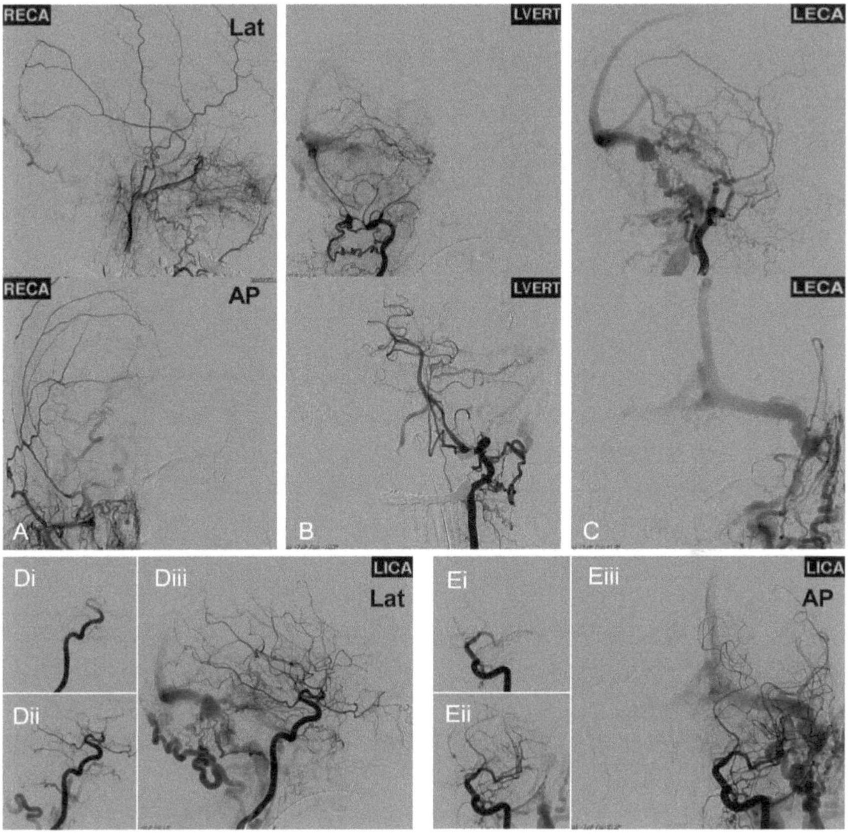


**Fig. 3** Case 3: Complex transverse sigmoid junction Type III dAVF. A 64-year-old woman presented with progressive confusion and falls, and an MRI demonstrated serpiginous vessels suspicious for a vascular malformation. A diagnostic cerebral angiogram demonstrated a complex left transverse sigmoid junction dAVF with arterial supply from the (A) right external carotid artery (ECA) via branch of the right occipital artery, (B) left vertebral artery via branches of the left posterior meningeal artery and muscular branches, (C) left ECA via branches of the left middle meningeal artery and occipital artery, and (D, E) left internal carotid artery (ICA) via the meningohypophyseal trunk (MHT). The MHT supply from the left ICA is demonstrated by time-lapse angiography images in lateral (Di–Diii) and anteroposterior views (Ei–Eiii). *AP*, anteroposterior; *Lat*, lateral; *LECA*, left external carotid artery; *LICA*, left internal carotid artery; *LVERT*, left vertebral artery; *RECA*, right external carotid artery.

meningeal artery and muscular branches (Fig. 3B), left MMA and occipital artery (Fig. 3C), and left meningohypophyseal trunk as demonstrated in time-lapse angiography still frames (Fig 3Di–Eiii). Evaluation of the angiographic runs determined that a joint transarterial and transvenous approach would be necessary to treat the dAVF.

A 5-Fr Envoy catheter was placed through the left common femoral venous sheath and placed into the left jugular bulb. An XT-27 microcatheter was placed over a Synchro$^2$ microwire (Stryker, Kalamazoo, MI) and navigated into the distal aspect of the left transverse sinus. A pressure manometer was connected to this catheter to record continuous venous pressure measurements throughout the transarterial portion of the embolization. Initial transvenous pressures in this fistulous pouch were elevated to 41 cm of $H_2O$, indicating arteriovenous shunting.

Through a 6-Fr Benchmark guide catheter placed in the left ECA, a Marathon microcatheter and Mirage microwire were navigated into the squamosal portion of the left MMA until just proximal to the fistulous point. Onyx 34 was injected until the majority of the fistulous point was embolized as it entered the left transverse sigmoid junction, taking care to prevent reflux along the petrous ridge to protect the facial nerve. Because the fistula appeared to drain directly into the transverse sigmoid sinuses with no distinct fistulous pouch for the majority of the fistulous connections (Fig. 4A), the venous sinus was also embolized.

Subsequent angiographic views of the right and left external carotid arteries demonstrated continued filling of the fistula by bilateral occipital branches across (Fig. 4B, C). The left vertebral artery also demonstrated persistent supply to the fistula via the posterior meningeal branches and occipital branches. The transvenous XT-27 which had been placed in the proximal left transverse sinus was used for additional embolization. First, the distal aspect of the transverse sinus was coiled, providing a scaffold for Onyx embolization (Fig. 4D). Transvenous embolization continued with further Onyx 34 injection until there was complete opacification of the sinus with no fistulous drainage from control injections of the left common carotid and vertebral arteries (Fig. 4E, F).

The patient was extubated and admitted to Neuro ICU where she was observed for 2 days before discharged to a skilled nursing facility. Her condition continued to improve such that 4 months later she was released from her assisted-living facility and returned to independent living.

## Conclusions

Transarterial embolization has proven to be a safe and effective strategy for managing high-grade dAVFs in large part due to the introduction of Onyx. Upon injection, Onyx forms a spongy plug that can fill the venous pouch from the arterial side, yielding high rates of obliteration with good

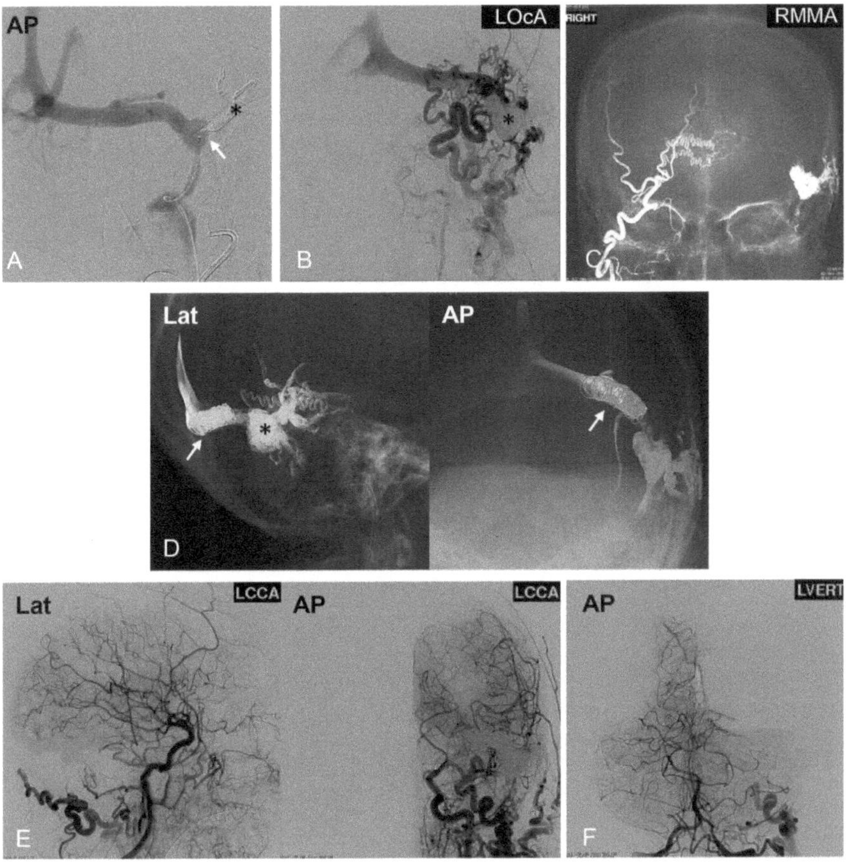


**Fig. 4** Case 3: Transarterial and transvenous embolization of transverse sigmoid junction Type III dAVF. A microcatheter was navigated into the squamosal portion of the left middle meningeal artery (MMA) until just proximal to the fistulous point, and Onyx 34 was injected until the majority of the fistulous points along the left transverse sigmoid junction, as well as multiple feeding vessels, were embolized (A, B). Care was taken to prevent reflux along the petrous ridge to protect the facial nerve. Angiographic views demonstrated continued filling of the fistula from bilateral external and occipital arteries draining directly into the transverse sigmoid sinuses without distinct fistulous pouch (B, C). The distal aspect of the transverse sinus was sequentially coiled off through a transvenous XT-27 microcatheter (D, *arrow*). Further transvenous embolization continued with Onyx 34 injection until there was complete opacification of the sinus (D, *star*) with no fistulous drainage from control injections via the left common carotid and vertebral arteries (E, F). *AP*, anteroposterior; *Lat*, lateral; *LCCA*, left common carotid artery; *LOcA*, left occipital artery; *LVERT*, left vertebral artery; *RMMA*, right middle meningeal artery.

long-term durability from a single intervention. Notably the risk of sinus occlusion with this strategy is quite low, minimizing the chance for a major complication during endovascular treatment. The transarterial approach can also be used in conjunction with other strategies, such as transvenous embolization, to safely treat complex lesions previously thought too dangerous for endovascular therapy.

## References


1. Halbach VV, Higashida RT, Hieshima GB, Mehringer CM, Hardin CW. Transvenous embolization of dural fistulas involving the transverse and sigmoid sinuses. *Am J Neuroradiol* 1989;**10**(2):385–92.
2. Kakarla UK, Deshmukh VR, Zabramski JM, Albuquerque FC, McDougall CG, Spetzler RF. Surgical treatment of high-risk intracranial dural arteriovenous fistulae: clinical outcomes and avoidance of complications. *Neurosurgery* 2007;**61**(3):447–59. https://doi.org/10.1227/01.NEU.0000280007.72190.F1.
3. Tomak PR, Cloft HJ, Kaga A, et al. Evolution of the management of tentorial dural arteriovenous malformations. *Neurosurgery* 2003;**52**(4):750–62. https://doi.org/10.1227/01.NEU.0000053221.22954.85.
4. Reynolds MR, Lanzino G, Zipfel GJ. Intracranial dural arteriovenous fistulae. *Stroke* 2017;**48**(5):1424–31. https://doi.org/10.1161/STROKEAHA.116.012784.
5. Natarajan SK, Ghodke B, Kim LJ, Hallam DK, Britz GW, Sekhar LN. Multimodality treatment of intracranial dural arteriovenous fistulas in the Onyx era: a single center experience. *World Neurosurg* 2010;**73**(4):365–79. https://doi.org/10.1016/j.wneu.2010.01.009.
6. Chandra R, Leslie-Mazwi T, Mehta B, et al. Transarterial onyx embolization of cranial dural arteriovenous fistulas: long-term follow-up. *Am J Neuroradiol* 2014;**35**(9):1793–7. https://doi.org/10.3174/ajnr.A3938.
7. Siekmann R. Basics and principles in the application of onyx LD liquid embolic system in the endovascular treatment of cerebral arteriovenous malformations. *Interv Neuroradiol* 2005;**11**(Suppl. 1):131–40. https://doi.org/10.1177/15910199050110s117.
8. Murayama Y, Viñuela F, Ulhoa A, et al. Nonadhesive liquid embolic agent for cerebral arteriovenous malformations: preliminary histopathological studies in swine rete mirabile. *Neurosurgery* 1998;**43**(5):1164–72. https://doi.org/10.1097/00006123-199811000-00081.
9. Rezende MTS, Piotin M, Mounayer C, Spelle L, Abud DG, Moret J. Dural arteriovenous fistula of the lesser sphenoid wing region treated with Onyx: technical note. *Neuroradiology* 2006;**48**(2):130–4. https://doi.org/10.1007/s00234-005-0020-9.
10. Borden JA, Wu JK, Shucart WA. A proposed classification for spinal and cranial dural arteriovenous fistulous malformations and implications for treatment. *J Neurosurg* 1995. https://doi.org/10.3171/jns.1995.82.2.0166.
11. Lekkhong E, Pongpech S, TerBrugge KG, et al. Arteriovenous shunts through occluded venous segments : experience in 51 patients. *Am J Neuroradiol* 2011;**32**:1738–44.
12. Lawton MT, Chun J, Wilson CB, Halbach VV. Ethmoidal dural arteriovenous fistulae: an assessment of surgical and endovascular management. *Neurosurgery* 1999;**45**(4):805–11. https://doi.org/10.1097/00006123-199910000-00014.
13. Doolan BJ, Paldor I, Mitchell PJ, Morokoff AP. First line direct access for transarterial embolization of a dural arteriovenous fistula: case report and literature review. *J Clin Neurosci* 2018;**48**:214–7. https://doi.org/10.1016/j.jocn.2017.10.070.

14. Kiyosue H, Hori Y, Okahara M, et al. Treatment of intracranial dural arteriovenous fistulas: current strategies based on location and hemodynamics, and alternative techniques of transcatheter embolization. *Radiographics* 2004;**24**(6):1637–53. https://doi.org/10.1148/rg.246045026.
15. Macdonald JHM, Millar JS, Barker CS. Endovascular treatment of cranial dural arteriovenous fistulae: a single-centre, 14-year experience and the impact of Onyx on local practise. *Neuroradiology* 2010;**52**(5):387–95. https://doi.org/10.1007/s00234-009-0620-x.
16. Nogueira RG, Dabus G, Rabinov JD, et al. Preliminary experience with Onyx embolization for the treatment of intracranial dural arteriovenous fistulas. *Am J Neuroradiol* 2008;**29**(1):91–7. https://doi.org/10.3174/ajnr.A0768.
17. Mantilla D, Le Corre M, Cagnazzo F, et al. Outcome of transarterial treatment of dural arteriovenous fistulas with direct or indirect cortical venous drainage. *J Neurointerv Surg* 2018;**10**(10):958–63. https://doi.org/10.1136/neurintsurg-2017-013476.
18. Abud TG, Nguyen A, Saint-Maurice JP, et al. The use of Onyx in different types of intracranial dural arteriovenous fistula. *Am J Neuroradiol* 2011;**32**(11):2185–91. https://doi.org/10.3174/ajnr.A2702.
19. Cognard C, Januel AC, Silva NA, Tall P. Endovascular treatment of intracranial dural arteriovenous fistulas with cortical venous drainage: new management using onyx. *Am J Neuroradiol* 2008;**29**(2):235–41. https://doi.org/10.3174/ajnr.A0817.
20. Elhammady MS, Wolfe SQ, Farhat H, Moftakhar R, Aziz-Sultan MA. Onyx embolization of carotid-cavernous fistulas: clinical article. *J Neurosurg* 2010;**112**(3):589–94. https://doi.org/10.3171/2009.6.JNS09132.
21. Micro Therapeutics. *Onyx liquid embolic system—instructions for use.* Dublin, Ireland: Medtronic; 2003. https://www.accessdata.fda.gov/cdrh_docs/pdf3/P030004c.pdf. [Accessed 18 November 2020].
22. Chaloupka JC, Huddle DC, Alderman J, Fink S, Hammond R, Vinters HV. A reexamination of the angiotoxicity of superselective injection of DMSO in the swine rete embolization model. *Am J Neuroradiol* 1999;**20**(3):401–10.
23. Natarajan SK, Britz GW. Multimodality treatment of brain arteriovenous malformations with microsurgery after emobolization with Onyx: single-center experience and technical nuances. *Neurosurgery* 2008;**62**(6):1213–26. https://doi.org/10.1227/01.NEU.0000316860.35705.AA.
24. Akin ED, Perkins E, Ross IB. Surgical handling characteristics of an ethylene vinyl alcohol copolymer compared with N-butyl cyanoacrylate used for embolization of vessels in an arteriovenous malformation resection model in swine. *J Neurosurg* 2003;**98**(2):366–70. https://doi.org/10.3171/jns.2003.98.2.0366.
25. Mounayer C, Hammami N, Piotin M, et al. Nidal embolization of brain arteriovenous malformations using onyx in 94 patients. *Am J Neuroradiol* 2007;**28**(3):518–23.
26. Tomsick TA, Purdy P, Horowitz M, et al. N-butyl cyanoacrylate embolization of cerebral arteriovenous malformations: results of a prospective, randomized, multi-center trial. *Am J Neuroradiol* 2002;**23**(5):748–55.
27. Loh Y, Duckwiler GR. A prospective, multicenter, randomized trial of the Onyx liquid embolic system and N-butyl cyanoacrylate embolization of cerebral arteriovenous malformations: clinical article. *J Neurosurg* 2010;**113**(4):733–41. https://doi.org/10.3171/2010.3.JNS09370.
28. Timasheff S, Bier M, Nord F. Aggregation phenomena in polyvinyl alcohol-acetate solutions. *Proc Natl Acad Sci* 1949;**35**:364–8.
29. Carlson AP, Taylor CL, Yonas H. Treatment of dural arteriovenous fistula using ethylene vinyl alcohol (Onyx) arterial embolization as the primary modality: short-term results. *J Neurosurg* 2007;**107**(6):1120–5. https://doi.org/10.3171/jns.2007.107.6.1120.
30. Gross BA, Albuquerque FC, Moon K, Mcdougall CG. Evolution of treatment and a detailed analysis of occlusion, recurrence, and clinical outcomes in an endovascular

library of 260 dural arteriovenous fistulas. *J Neurosurg* 2017;**126**(6):1884–93. https://doi.org/10.3171/2016.5.JNS16331.

31. Chew J, Weill A, Guilbert F, Raymond J, Audet ME, Roy D. Arterial onyx embolisation of intracranial DAVFs with cortical venous drainage. *Can J Neurol Sci* 2009;**36**(2):168–75. https://doi.org/10.1017/S0317167100006521.
32. Chong WKW, Holt M. Endovascular therapy for intracranial dural arteriovenous fistulas. *Neuroradiol J* 2006;**19**(4):537–49. https://doi.org/10.1177/197140090601900411.
33. Hu YC, Newman CB, Dashti SR, Albuquerque FC, McDougall CG. Cranial dural arteriovenous fistula: transarterial Onyx embolization experience and technical nuances. *J Neurointerv Surg* 2011;**3**(1):5–13. https://doi.org/10.1136/jnis.2010.003707.
34. Ertl L, Brückmann H, Kunz M. Endovascular therapy of low- and intermediate-grade intracranial lateral dural arteriovenous fistulas: a detailed analysis of primary success rates, complication rates, and long-term follow-up of different technical approaches. *J Neurosurg* 2017;**126**:360–7. https://doi.org/10.3171/2016.2.JNS152081.360.
35. Ambekar S, Gaynor BG, Peterson EC, Elhammady MS. Long-term angiographic results of endovascularly "cured" intracranial dural arteriovenous fitulas. *J Neurosurg* 2016;**124**(4):1123–7. https://doi.org/10.3171/2015.3.JNS1558.
36. Ng PP, Higashida RT, Cullen S, Malek R, Halbach VV, Dowd CF. Endovascular strategies for carotid cavernous and intracerebral dural arteriovenous fistulas. *Neurosurg Focus* 2003;**15**(4):1–6. https://doi.org/10.3171/foc.2003.15.4.7.
37. Ghobrial GM, Marchan E, Nair AK, et al. Dural arteriovenous fistulas: a review of the literature and a presentation of a single institution's experience. *World Neurosurg* 2013;**80**(1–2):94–102. https://doi.org/10.1016/j.wneu.2012.01.053.
38. Rangel-Castilla L, Barber SM, Klucznik R, Diaz O. Mid and long term outcomes of dural arteriovenous fistula endovascular management with Onyx. Experience of a single tertiary center. *J Neurointerv Surg* 2014;**6**(8):607–13. https://doi.org/10.1136/neurintsurg-2013-010894.
39. Lv X, Jiang C, Li Y, Wu Z. Results and complications of transarterial embolization of intracranial dural arteriovenous fistulas using Onyx-18: clinical article. *J Neurosurg* 2008;**109**(6):1083–90. https://doi.org/10.3171/JNS.2008.109.12.1083.
40. Geibprasert S, Pongpech S, Armstrong D, Krings T. Dangerous extracranial-intracranial anastomoses and supply to the cranial nerves: vessels the neurointerventionalist needs to know. *Am J Neuroradiol* 2009;**30**(8):1459–68. https://doi.org/10.3174/ajnr.A1500.

CHAPTER 6

# Transvenous embolization of cerebral dural arteriovenous fistulas

**Jawad M. Khalifeh, Robert T. Wicks, Jennifer E. Kim, Justin M. Caplan, and Cameron G. McDougall**
Department of Neurosurgery, Johns Hopkins University School of Medicine, Baltimore, MD, United States

## Introduction

The natural history and indications for treatment of dural arteriovenous fistulas (dAVFs) are covered elsewhere in this volume. This chapter focuses on transvenous approaches in the treatment of dAVFs. Although transvenous approaches are particularly helpful in the treatment of indirect carotid cavernous fistulas (CCFs), this pathology is addressed separately in this book.

Endovascular embolization is the primary treatment for most dAVFs.[1–3] The goal of endovascular therapy is the occlusion of the fistula pouch and its proximal venous drainage.[1] Complete obliteration of the fistulous connection eliminates the pathologic arteriovenous shunt and, if present, the corresponding venous hypertension, thereby alleviating neurologic symptoms and reducing the long-term risk of intracranial hemorrhage.[2,4] Importantly, incomplete arterial or venous embolization is not a durable solution. It is associated with reconstitution of the dAVF via the recruitment of arterial collaterals, and the potential development of high-risk leptomeningeal or cortical venous drainage over the course of weeks to months.[5,6]

Transvenous embolization is particularly effective in the treatment of large or complex dAVFs with multiple arterial feeders and in incompletely treated dAVFs that are not readily or safely accessible by transarterial routes.[1,2,7,8] This chapter reviews the indications and procedural technique for transvenous embolization of cerebral dAVFs. We will discuss important anatomical considerations and present illustrative cases showing both straightforward and more challenging scenarios.

*Cerebral Dural Arteriovenous Fistulas*
https://doi.org/10.1016/B978-0-12-819525-3.00006-X

## Historical perspective

Arteriovenous communications or "aneurysmal varices" were first described by Dr. William Hunter in the mid-18th century, in a series of remarks on aortic and peripheral aneurysms.[9] He noted in his observations that "the vein will be dilated or become varicose, and it will have a pulsatile jarring motion on account of the stream from the artery" and recommended non-operative treatment of these malformations as he felt that the rupture risk was low as compared with "true aneurysms."[10] In 1873 Francesco Rizoli described a postmortem finding of a dural-based arteriovenous aneurysm and an abnormal communication between hypertrophied branches of the occipital artery and transverse sinus.[11] Harvey Cushing and Percival Bailey, as well as Walter Dandy published their series of intracranial arteriovenous "aneurysms" many of which were CCFs and were attributed to trauma.[12–14]

Surgical management of these lesions ranged from primary ligation of the common, internal, and/or external carotid arteries, trapping of the fistula by ligating the cervical and intracranial carotid in the case of CCFs, to obliteration with muscle or other injected particulates into the feeding artery.[13, 15, 16] Through most of the 20th century, microsurgical resection or disconnection of draining veins was the mainstay of treatment for symptomatic dAVFs. Hugosson, followed by Sundt and Piepgras were the first to describe microsurgical excision of the transverse sinus to occlude transverse-sigmoid sinus dAVFs.[17, 18] However, these procedures were often complicated by high blood loss, high rates of recurrence, and postoperative morbidity and mortality. The advent of digital subtraction angiography and modern endovascular techniques revolutionized the treatment of dAVFs, allowing for direct access to the lesion and its feeding arteries and draining veins, as well as transcatheter delivery of balloons, particles, liquid adhesives, and coils and other occlusive agents.

Early experience with transvenous embolization demonstrated high cure rates with dAVFs of the transverse, sigmoid, and cavernous sinuses.[19, 20] Studies of transvenous embolization report 71%–87.5% angiographic cures and 83%–96% clinical cures.[21] However, the feasibility of a transvenous approach may be limited by anatomic constraints, such as thrombosis along the involved dural venous sinuses, narrow or tortuous venous anatomy, or the need to maintain normal venous drainage pathways.[22] As such, the transarterial approach has become the preferred method of endovascular access, owing to its relative safety, efficacy, and familiarity in navigation and technique. The increased use of the transarterial approach has been greatly facilitated by the recent availability of nonadhesive liquid embolic agents capable of distal penetration into the venous outflow pouch.[7, 23, 24]

## Overview and indications

Transvenous embolization consists of the selective retrograde catheterization of a draining dural venous sinus or cortical vein to access the fistulous zone.[2, 7] Embolic materials are typically delivered through the distal microcatheter tip to occlude the proximal side of the venous outflow pouch.[7] In some cases, nonocclusive or remodeling techniques using angioplasty and stent deployment have been reported for dAVFs of the transverse and sigmoid sinus.[25] The aim of treatment is the complete obliteration of the dAVF with preservation of normal venous drainage pathways. Embolic materials may include detachable coils, solid particulates (e.g., polyvinyl alcohol particles, silk sutures, and poly-L-lactic acid microspheres), and/or liquid agents (e.g., ethyl alcohols, *n*-butyl cyanoacrylate (*n*-BCA), ethylene vinyl alcohol copolymer) (Onyx, Medtronic).[4, 24]

Transvenous embolization is performed safely when the involved draining veins or dural venous sinuses do not contribute to normal venous drainage as they are rendered nonfunctional by the dAVF.[7, 8, 26] The involved dural venous sinuses can then be deliberately occluded as the cerebral venous outflow is directed through other unaffected or collateral routes. Occasionally, endovascular transvenous access to the involved dural venous sinus is obstructed by stenosis, occlusion, or compartmentalization within the sinus. Thrombosis or stenosis of the involved dAVF sinus can be observed in 71% and 25% of patients, respectively.[27] Transcranial surgical approaches may be required to adequately cure the fistula,[7, 28] although, it is possible to navigate through a partially or completely thrombosed dural venous sinus using endovascular recanalization techniques.[29, 30]

Selection of dAVFs for transvenous embolization requires accurate localization of the fistulous connection, patterns of venous outflow, and anticipated changes in venous drainage following embolization. Complications of venous infarction and intracranial hemorrhage may occur following embolization of a dural venous sinus segment that contributes to normal venous drainage. Thus the observation of prominent antegrade cortical venous flow into the diseased dural venous sinus segment (e.g., vein of Labbé into the transverse sinus and vein of Trolard into the superior sagittal sinus) should prompt reconsideration of alternative embolization approaches.[19] In addition, incomplete or partial embolization of a diseased dural venous sinus could lead to diversion of arteriovenous shunt flow into cortical veins. This may result in worsening of neurologic symptoms and an unfavorable conversion of the dAVF into an aggressive lesion that is prone to infarction and hemorrhage.[6,7] Angiographic balloon test occlusion can be a useful adjunct to evaluate the response to attempted dural venous sinus occlusion and all potential cortical venous reflux possibilities.[8,27]

## Techniques and nuances

Once the decision has been made to treat a dAVF, the key to successful treatment is a thorough understanding of the lesion. The precise location of the fistula must be determined. In some instances, there may be more than one site of arteriovenous shunting and this must be appreciated before treatment. As fistulas can be large with multiple arterial inputs, the initial site of arteriovenous shunting can be difficulty to perceive. Most often superselective catheterization is helpful, looking on the initial frames to determine where the transition to the vein is first seen. This is greatly facilitated by using general anesthesia so that there is no motion artifact to impair image quality. Occasionally, multiple superselective pedicle catheterizations are required to ascertain with certainty that there is a common point of arteriovenous shunting, and not multiple fistulas. When a fistula has recruited arterial supply from a wide variety of branches, it is sometimes helpful to look on the side contralateral to the fistula as the lesser supply from the contralateral side can more easily reveal the precise location of the fistula. When flow is very rapid, partial embolization of some of the feeding pedicles may allow the fistula site to be more clearly seen. Additionally, transarterial embolization that slows the flow through the fistula can facilitate the ultimate transvenous occlusion of the fistula.

Once the site of the fistula is confirmed, the next question is how best to get to the site of the fistula. The pathways may be arterial, venous, direct puncture, open surgical, or some combination of these. Usually the easiest path is also the safest. Whatever pathway is chosen, the closer one is to the fistula and the venous outflow, the higher the probability of curing the fistula. With respect to endovascular approaches, the arterial pathway has the advantage of easier closure of the transit vessel should a perforation occur, but it is not always possible to get close enough to the fistula to safely deliver the embolic agent in a quantity sufficient to occlude the fistulous pouch and venous outflow. Additionally, when there are multiple arterial inflow pedicles to the fistula, it may be difficulty to prevent the embolic agent from being swept through the fistula by these competing arterial inputs. Additionally the risk of nontarget embolization is invariably higher from a transarterial position. Conversely, with the transvenous approach, perforation of the catheterized (arterialized) vein can be difficult to manage because the catheter is likely to be on the downstream side of the perforation, meaning that control of the arterialized inflow—and extravasation—is beyond the tip of the catheter.[31] In favor of the venous approach, the risk of nontarget embolization is very low. Once the fistula site is reached transvenously, arrival at the fistula site can be confirmed with a microcatheter injection, which will frequently show retrograde filling of the feeding arteries. Once appropriate catheter position is achieved and confirmed, there is minimal risk in taking the outflow vein as the arterialized vein is providing no functional venous drainage. Indeed, reliable, durable cure of the fistula requires the occlusion of the venous side of the fistula. In some instances, fistulas can be cured taking the sole outflow path of the fistula at some distance from the fistula itself, but the farther from the fistula that this is performed, the more likely it is that the fistula will find an alternative venous outflow and thereby not be cured.

## Clinical cases

To illustrate techniques and technical nuances in some detail, three illustrative cases are presented.

### Case 1

An otherwise healthy middle-aged gentleman presented with left-sided pulsatile tinnitus. A left transverse/sigmoid Borden Type 1 fistula was identified (Fig. 1A). Although the patient understood that the fistula did not present significant risk in terms of hemorrhage the symptoms were persistent and sufficiently disturbing that he wished to undergo treatment. Transvenous embolization was chosen since cure of the fistula via transarterial embolization was unlikely given the extent of the arterial feeding vessels (Fig. 1B). In some similar fistulas, it may be possible to use a large compliant balloon to protect the sigmoid sinus during transarterial embolization with liquid embolic,[32] but that was not thought to be a viable option for this patient. Similarly, in some patients a separate, loculated, fistulized channel exists in the sigmoid sinus that can be transvenously (or transarterially) embolized while preserving the normal portion of the sinus[33, 34] but no such channel could be identified in this patient. On demonstrating that the left vein of Labbé drained across the torcula to the right and that the arterial inflow did not extend medially as far as the vein of Labbé, transvenous embolization and sacrifice of the sigmoid sinus was determined to be a viable option for cure of the fistula (Fig. 1C–E). Because the site of arterial inflow was very close to the inflow point of Labbé and because occluding vascular structures at a precise point with coils can be difficult, a "barrier" was established just lateral to the vein of Labbé by placing a microvascular plug (MVP, Medtronic). This effectively occluded the sinus at the transverse sigmoid junction and provided a backstop against which coils could be more densely packed without placing the vein of Labbé at risk (Fig. 1F and G). Additionally, before the venous embolization, the occipital artery fistula inflow was embolized using Onyx (Medtronic) to facilitate the subsequent venous embolization thrombosis (Fig. 1H–K).

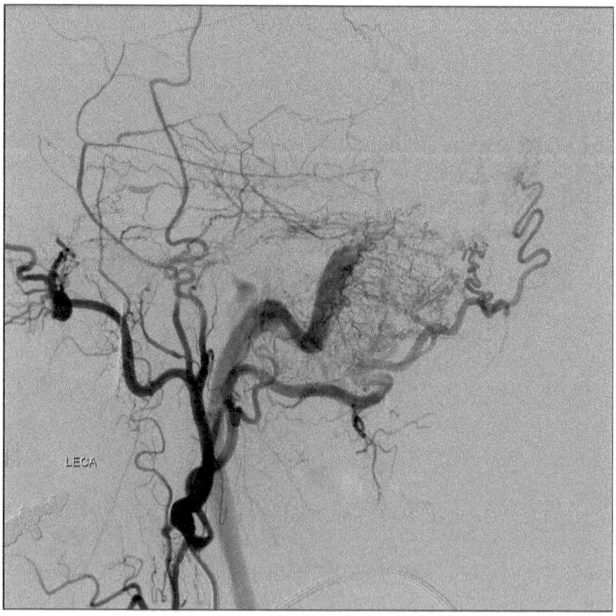


(A)

**Fig. 1** (A) Lateral view of a left external carotid artery injection. There is brisk arteriovenous shunting to the left sigmoid sinus with the arterial supply including multiple branches of the left middle meningeal artery and a markedly enlarged left occipital artery.

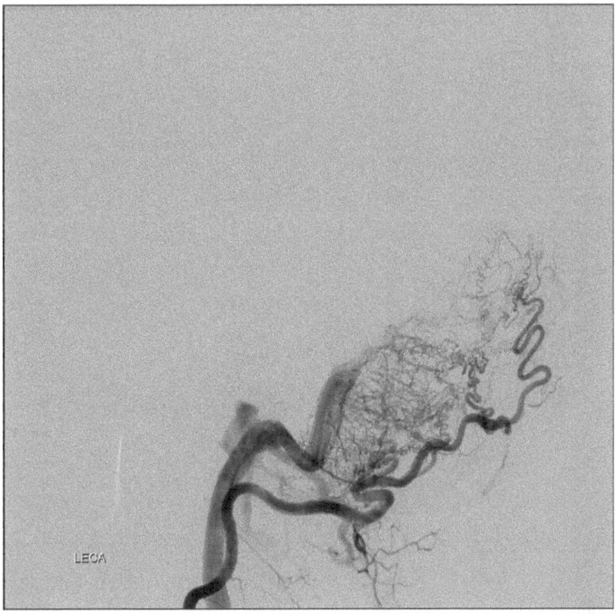


(B)

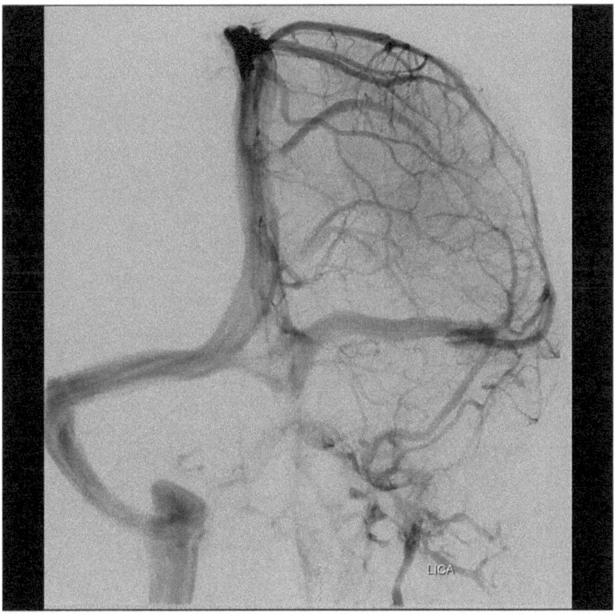


(C)

**Fig. 1, cont'd** (B) Lateral projection of a selective injection of the left occipital artery demonstrating marked hypertrophy, and a myriad of branches arising from the occipital artery to supply the fistula with no focal venous receptacle pouch. (C) Frontal projection of a left internal carotid artery injection, venous phase, with nonvisualization of the left sigmoid sinus, due to the high flow arterialization of the sigmoid sinus. The normal cortical venous drainage of the right hemisphere, including the vein of Labbé, drains to the left transverse sinus and then retrograde across the torcula to the right jugular vein.

*(Continued)*

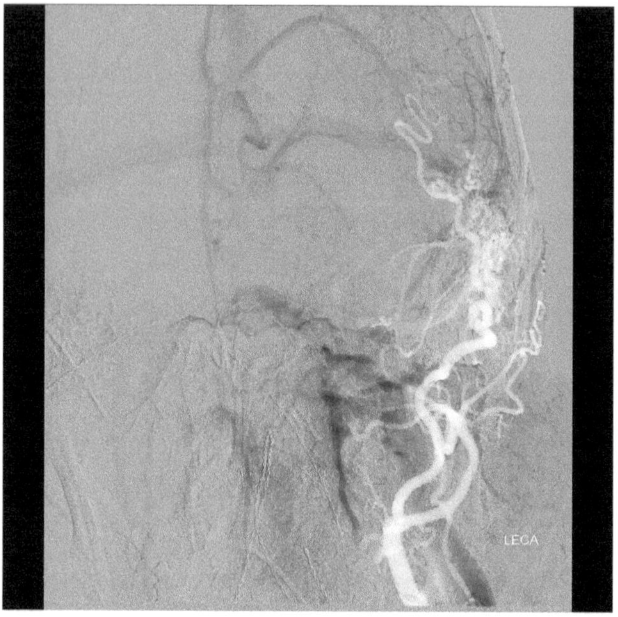


(D)

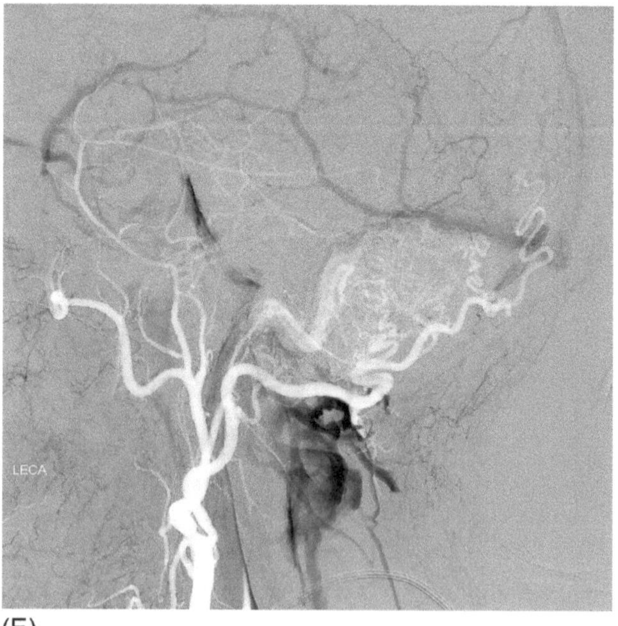


(E)

**Fig. 1, cont'd** (D and E) Frontal and lateral projections of a left common carotid artery injection. By manipulating the phase of the subtraction mask, moving it to the early arterial phase, it is possible to demonstrate the relationship of the vein of Labbé to the most medial portion of the pathological arterialized inflow. In this instance the vein of Labbé is quite close to the fistula. It is imperative that the vein of Labbé's connection to the transverse sinus and outflow to the torcula be preserved.

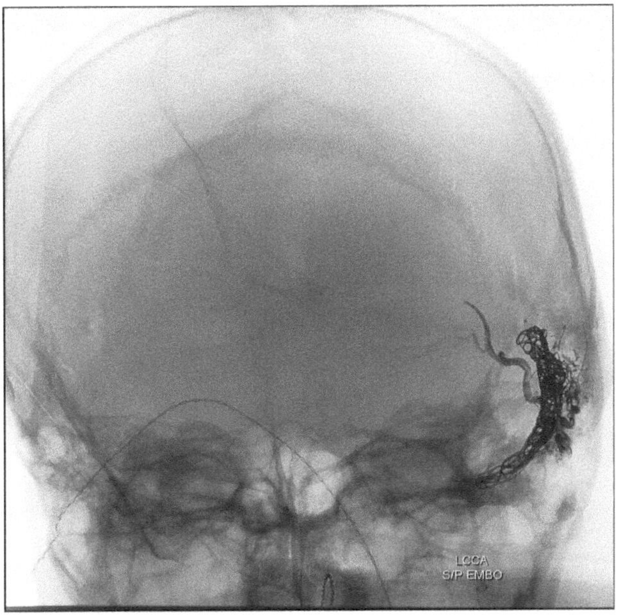


(F)

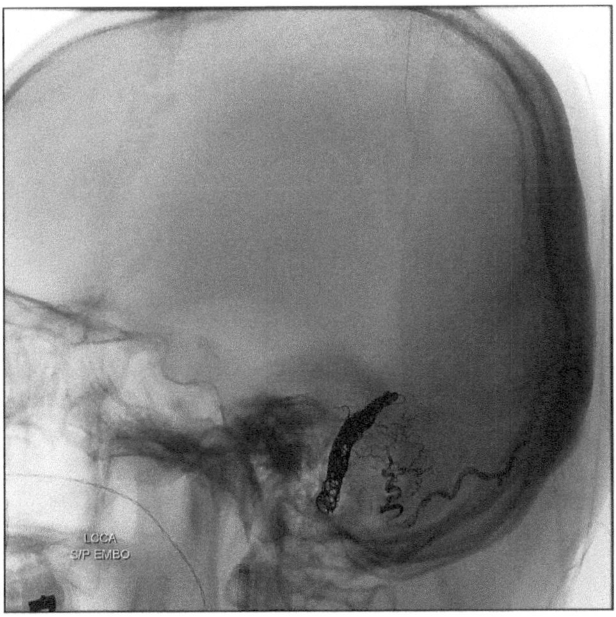


(G)

**Fig. 1, cont'd** (F and G) After embolization, unsubtracted frontal and lateral images. Visible are the embolic materials, including the arterially injected Onyx, and the transvenously placed MVP plug and coils.

*(Continued)*

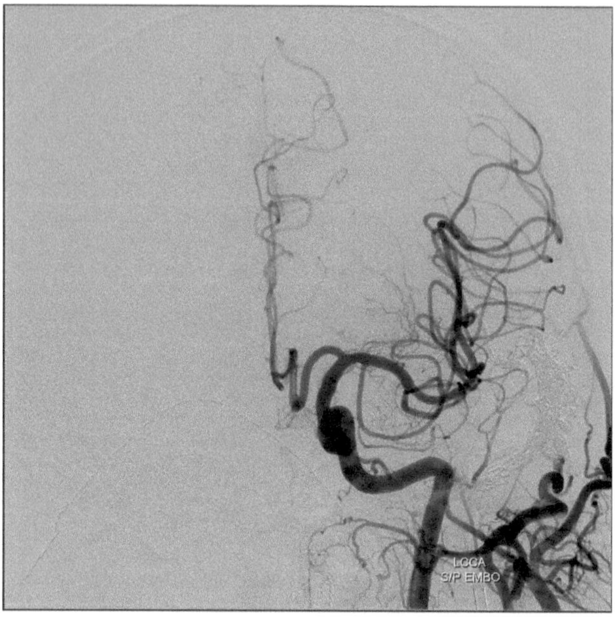


(H)

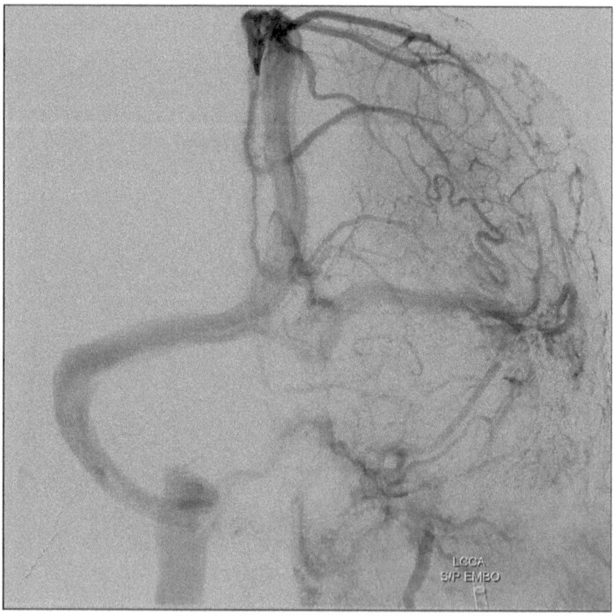


(I)

**Fig. 1, cont'd** (H–K) Left common carotid artery injection in the frontal projection (early phase (H); late phase (I)) and lateral projection (early phase (J); late phase (K)). Injections revealed a cure of the fistula, which was durable on follow-up angiography at 3 months.

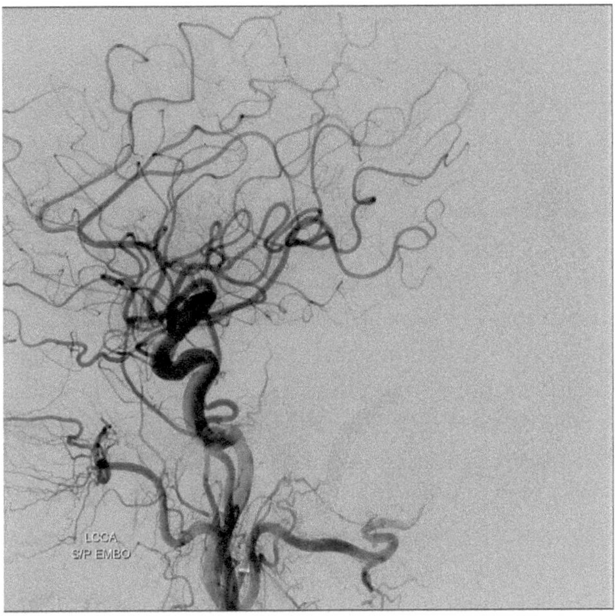


(J)

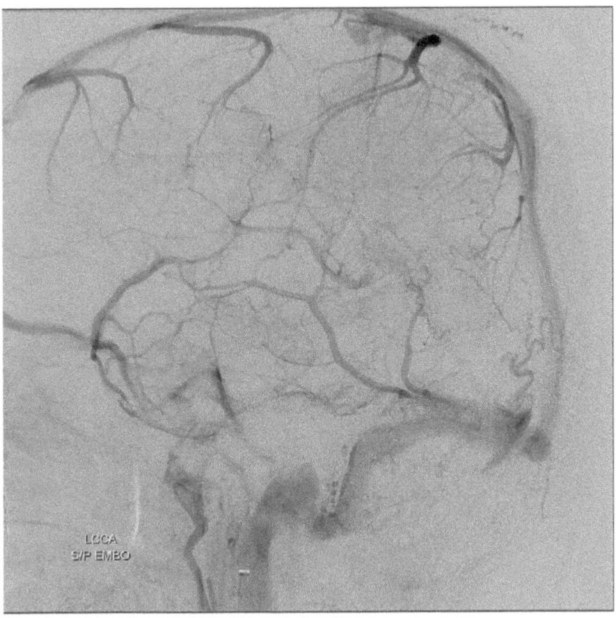


(K)

**Fig. 1, cont'd**

## Case 2

A 60-year-old gentleman presented after experiencing sinus infections of increasing frequency and was found to have a left frontal sinus mucocele. He had a history of surgical repair of a left orbital fracture secondary to an assault 30 years before presentation. On preoperative MR Brain for planned surgical repair of the sinus mucocele, a dilated vein along the floor of the medial left frontal fossa was identified concerning for underlying vascular malformation (Fig. 2A). He underwent diagnostic cerebral arteriogram confirming a Borden Type 3 ethmoidal dAVF primarily supplied by ethmoidal branches of the left ophthalmic artery and maxillary artery and draining to an anterior cortical vein that then entered the anterior superior sagittal sinus (Fig. 2B and C). Selective injection of the left maxillary artery did not identify a safe, direct transarterial route to the fistula (Fig. 2D). While many ethmoidal dAVFs are managed with open surgery, the straightforward venous access in this case suggested a safe endovascular alternative. An intermediate catheter within the posterior one-third of the transverse sinus allowed for sufficient support to place a Headway Duo (Microvention, Inc.) microcatheter into the anterior aspect of the superior sagittal sinus to access the fistulous point (Fig. 2E and F). Coils were placed into this fistulous sac resulting in complete occlusion (Fig. 2G and H).

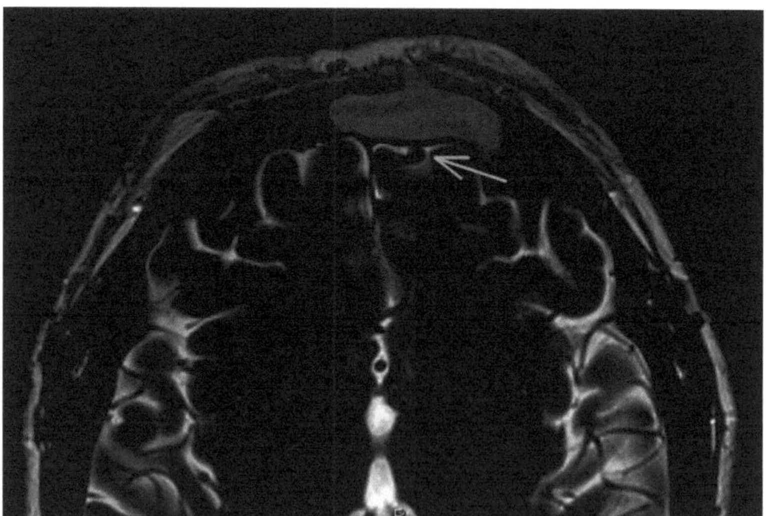

(A)

**Fig. 2** (A) Axial MR Brain T2 sequence without contrast obtained to evaluate left frontal sinus mucocele. *Arrow* denotes dilated cortical vein concerning for underlying vascular malformation.

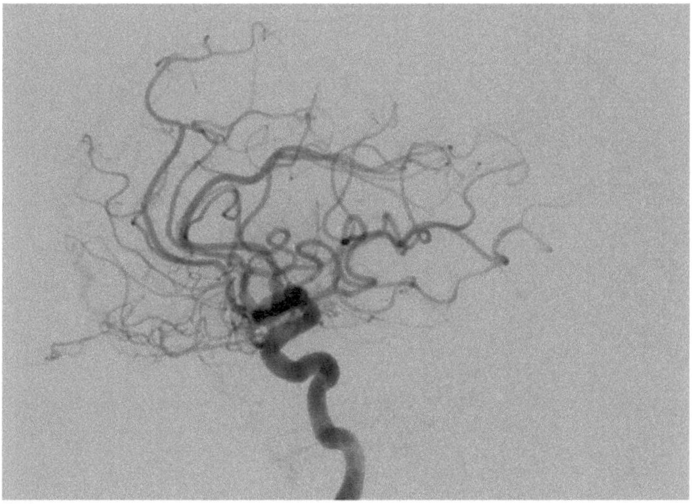

(B)

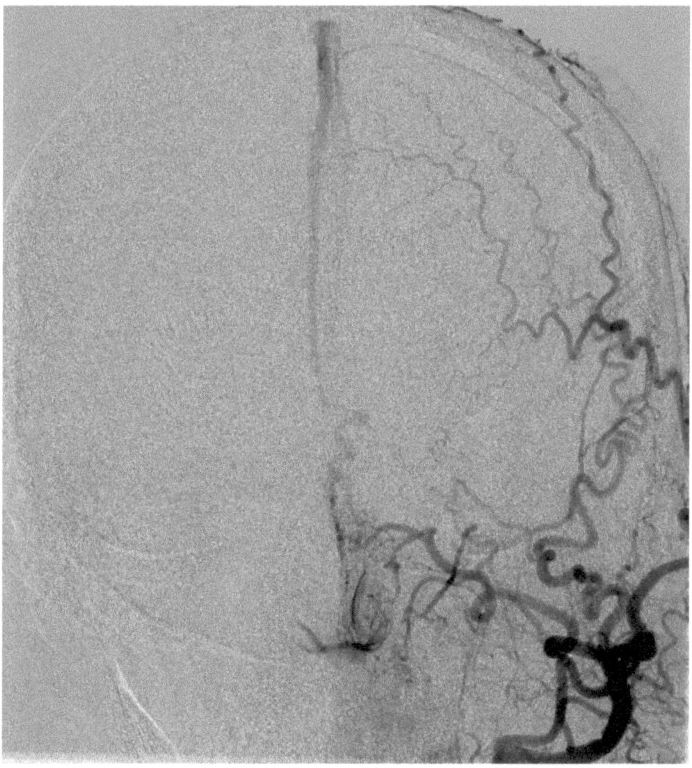

(C)

**Fig. 2, cont'd** (B) Lateral projection of left internal carotid injection revealing anterior cranial fossa early cortical venous drainage via ethmoidal branches of the left ophthalmic artery. (C) Posterior-anterior projection of left external carotid injection shows hypertrophied terminal maxillary branches supplying the anterior cranial fossa dAVF draining via anterior frontal cortical vein to the anterior superior sagittal sinus.

*(Continued)*

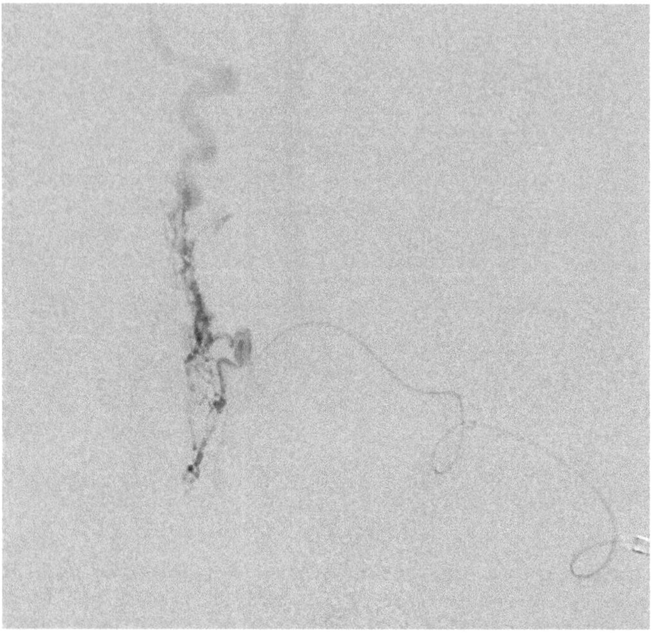

(D)

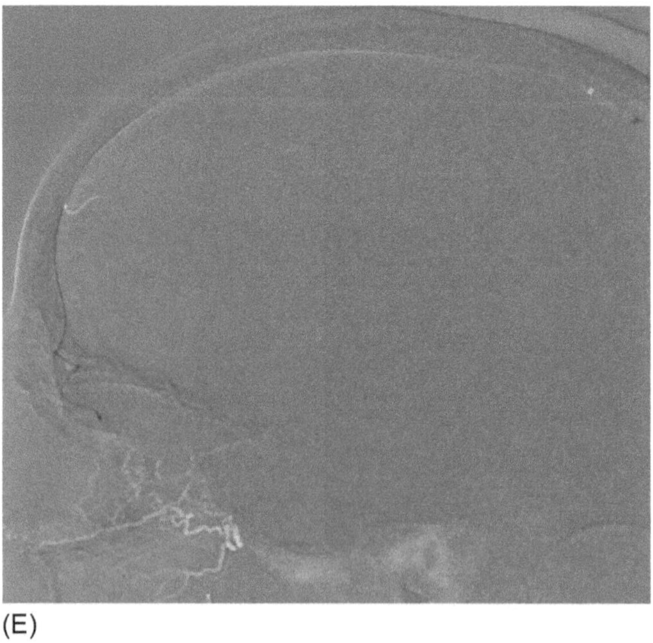

(E)

**Fig. 2, cont'd** (D) Selective microcatheterization of distal left maxillary artery branches supplying the anterior fossa dAVF. No direct route for transarterial embolization is identified without risk to the ophthalmic artery given ethmoidal anastomoses. (E) Lateral projection roadmap image revealing the subtracted arterial supply to the fistula. The intermediate catheter is within the posterior one-third of the superior sagittal sinus to the left with microcatheterization of the anterior cranial fossa fistula.

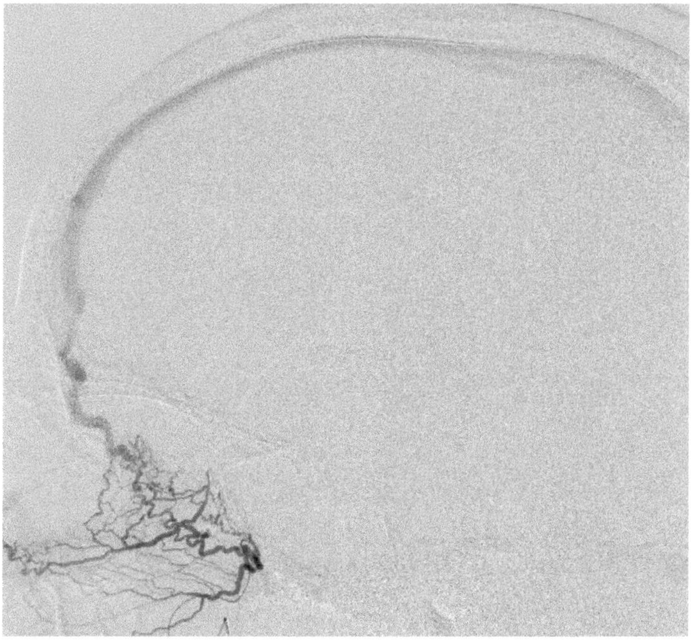

(F)

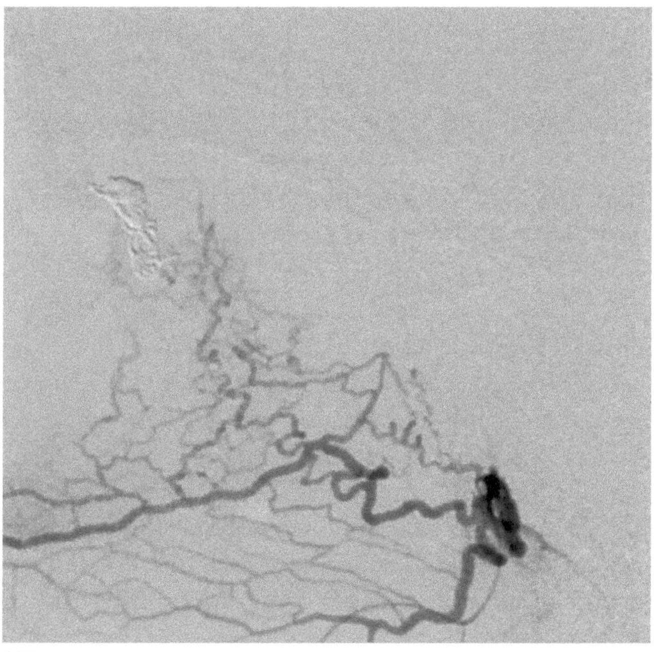

(G)

**Fig. 2, cont'd** (F) Lateral projection with distal left maxillary artery injection confirming location of the anterior cranial fossa dAVF pouch. (G) Magnified lateral projection of distal left maxillary artery injection showing successful coil embolization of the anterior fossa dAVF with no further early venous drainage.

*(Continued)*

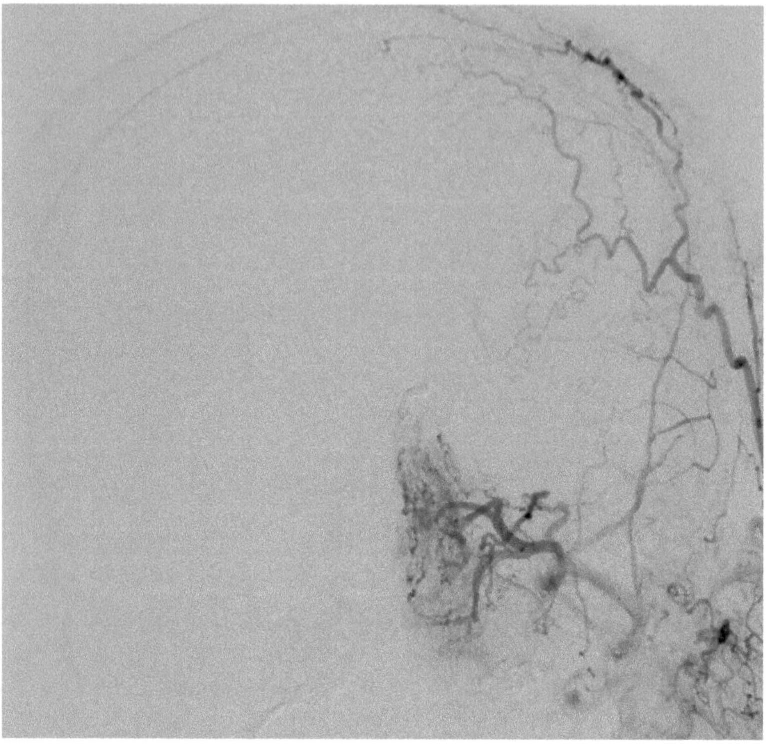

(H)

**Fig. 2, cont'd** (H) Posterior-anterior projection of left external carotid artery injection confirming successful coil embolization of the anterior fossa dAVF.

## Case 3

A septuagenarian presented with a mild right hemisphere stroke from which she completely recovered. The stroke was ultimately attributed to paroxysmal atrial fibrillation. Investigations related to her stroke revealed a left hemispheric vascular abnormality consistent with a left middle fossa dAVF. The fistula, though asymptomatic, was deemed to be of significant risk secondary to the greatly dilated cortical venous varices which drained ultimately via the left vein of Labbé. Because of comorbidities, the patient was judged to be of increased risk for open surgical treatment and was referred for endovascular therapy.

The arterial supply to the fistula came mainly from the left ophthalmic and middle meningeal arteries, both of which were markedly hypertrophied, and to a lesser extent from additional external carotid artery branches (Fig. 3A and B). Transarterial embolization via the middle meningeal artery was planned but the access was limited by proximal tortuosity

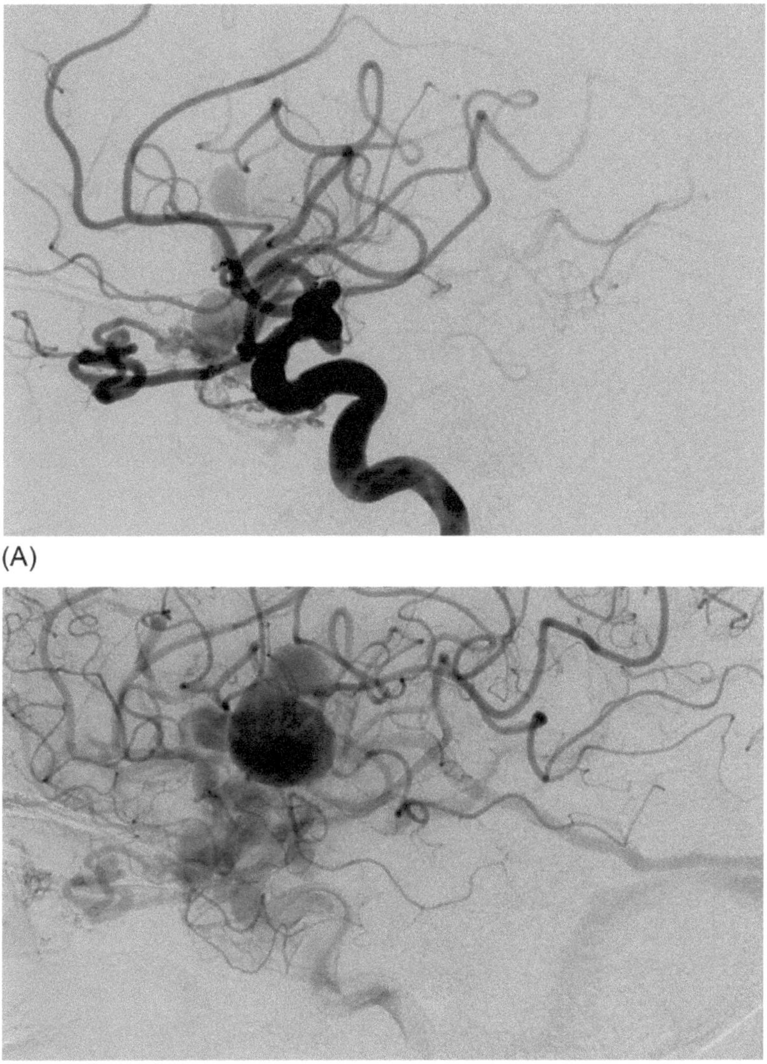

(A)

(B)

**Fig. 3** (A and B) Left internal carotid injections, lateral projection in the early (A) and mid-arterial (B) phases. Note the enlarged ophthalmic artery, which was not suitable for access to the fistula.

*(Continued)*

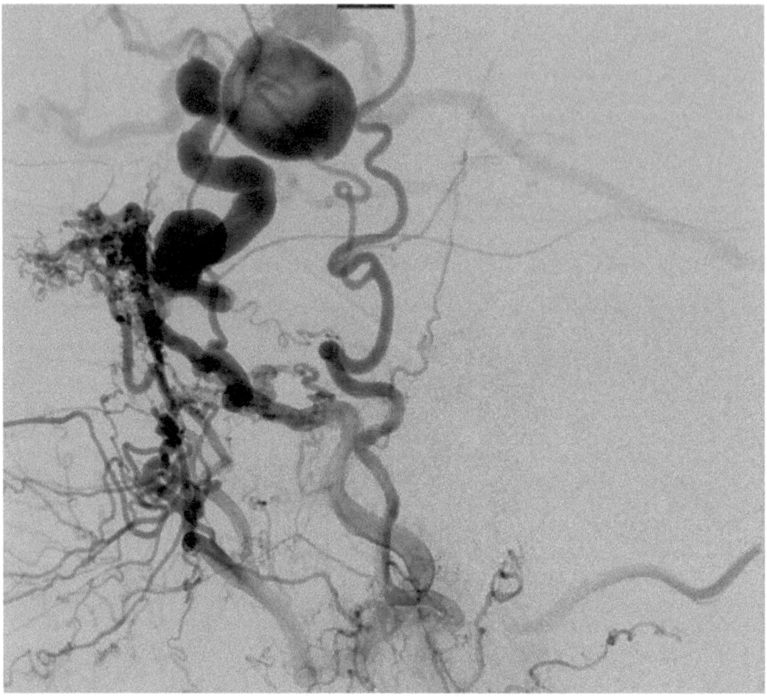

(C)

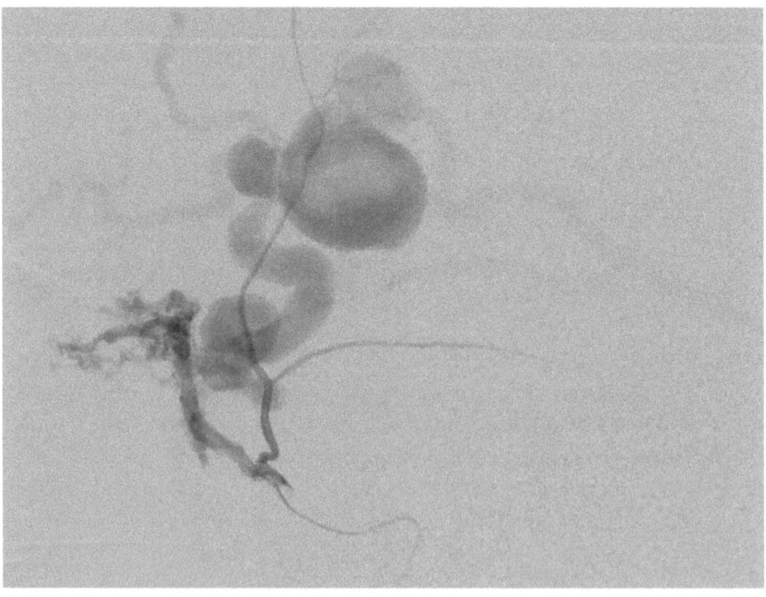

(D)

**Fig. 3, cont'd** (C) Lateral projection of the left external carotid artery injection with markedly enlarged middle meningeal artery supplying the fistula. (D) Lateral projection of selective catheterization of the left middle meningeal artery showing the fistulous pouch.

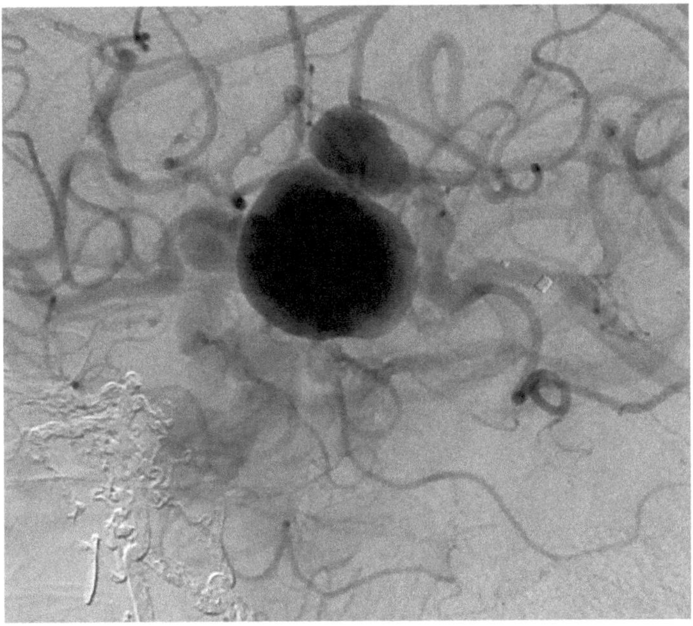

(E)

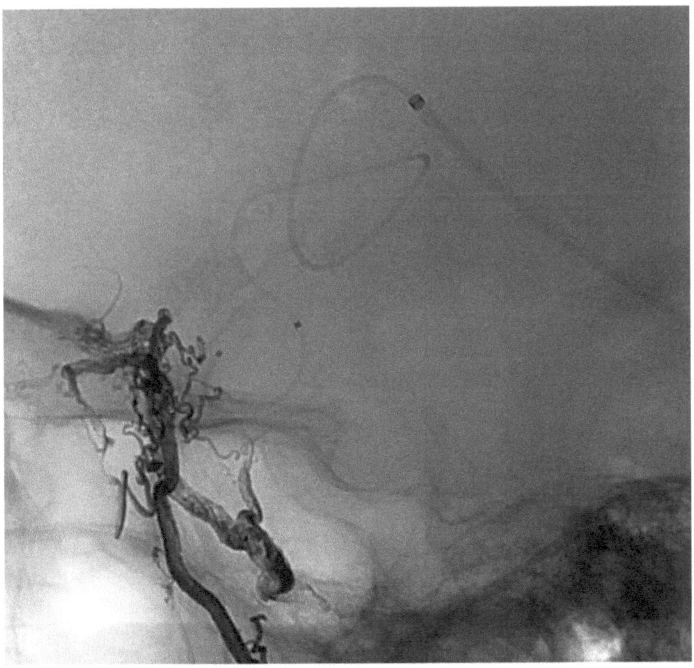

(F)

**Fig. 3, cont'd** (E and F) Transvenous approach to the middle fossa dAVF 2 days after attempted transarterial embolization via the middle meningeal artery. Lateral projection unsubtracted view (E) and subtracted view (F). Note intermediate catheter in vein of Labbé and microcatheter tip at the fistula.

*(Continued)*

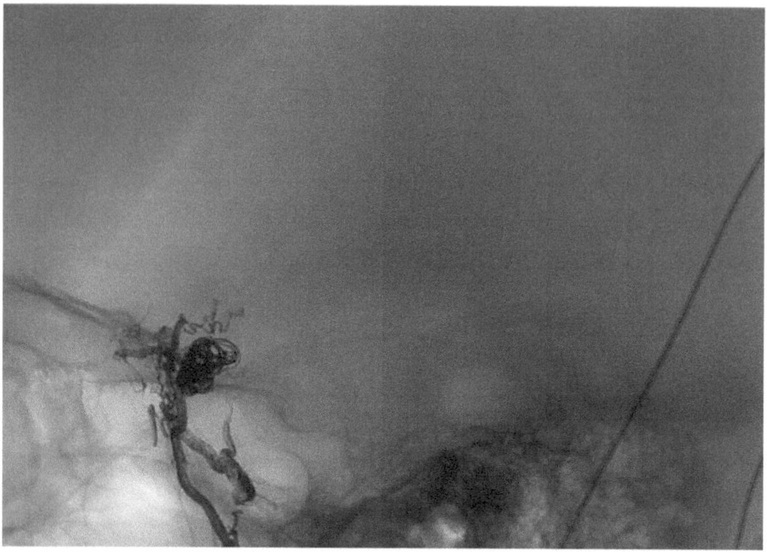

(G)

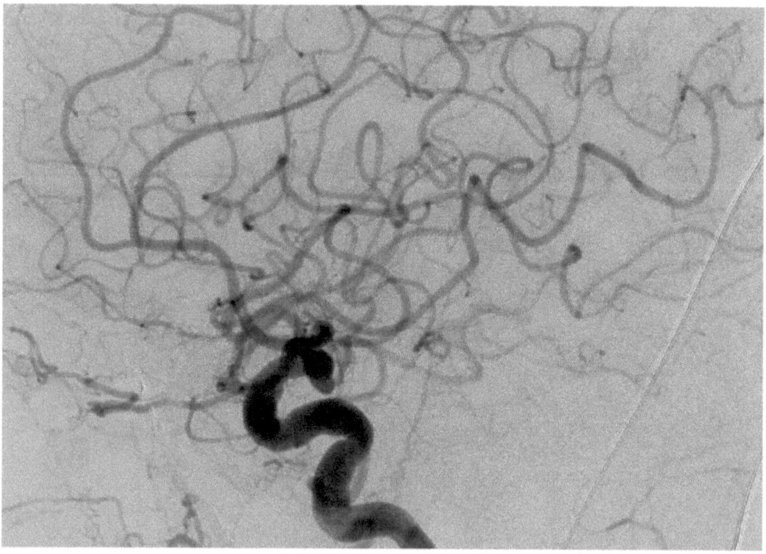

(H)

**Fig. 3, cont'd** (G–I) Left common carotid artery injection in the lateral projection 3 months after treatment. No residual or recurrent fistula. Note the coils at the site of the fistula (G), with no early arterial filling on arterial phase (H) and continued patency of the left vein of Labbé on the venous phase (I).

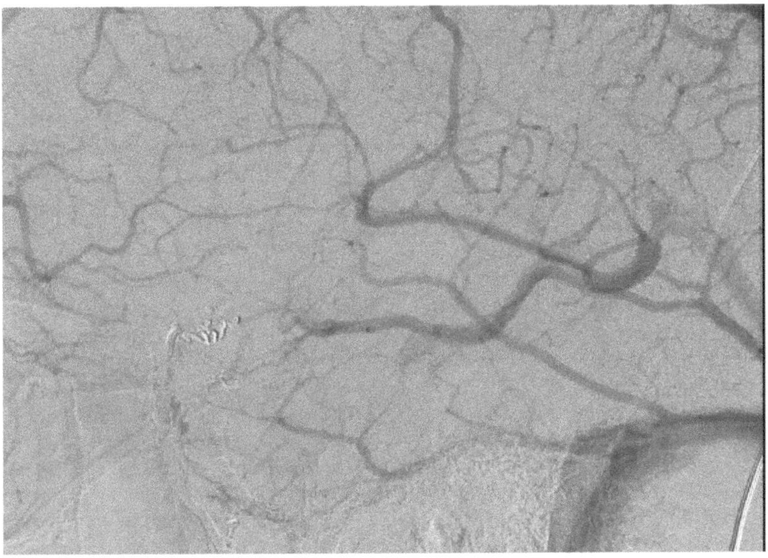

(I)

**Fig. 3, cont'd**

(Fig. 3C and D. Despite using a Scepter balloon (Microvention, Inc.) in the middle meningeal artery to minimize reflux of the embolisate, the attempt was never-the-less limited by proximal reflux in the middle meningeal artery. With limited penetration into the fistulous point, the remaining arterial supply continued to fill the fistula. Two days later the patient underwent treatment via a transvenous approach to access the fistula via the vein of Labbé (Fig. 3E). Access to the fistula was difficult because of the marked tortuosity of the venous drainage but placement of an intermediate catheter in the vein of Labbé assisted in the access. Ultimately the site of the fistula was reached, and the fistula was embolized with coils (Fig. 3F and G). Three months after embolization angiographic follow-up showed that the fistula remained occluded. The left vein of Labbé remained patent and the patient remained neurologically intact. (Fig. 3H and I).

## References


1. Miller TR, Gandhi D. Intracranial dural arteriovenous fistulae: clinical presentation and management strategies. *Stroke* 2015;**46**(7):2017–25.
2. Gandhi D, Chen J, Pearl M, Huang J, Gemmete JJ, Kathuria S. Intracranial dural arteriovenous fistulas: classification, imaging findings, and treatment. *AJNR Am J Neuroradiol* 2012;**33**(6):1007–13.

3. Rammos S, Bortolotti C, Lanzino G. Endovascular management of intracranial dural arteriovenous fistulae. *Neurosurg Clin N Am* 2014;**25**(3):539–49.
4. Grossberg J, Cawley C. Treatment of other intracranial dural arteriovenous fistulas. In: *Youmans neurological surgery*. Philadelphia, PA: Saunders; 2017.
5. van Dijk JM, terBrugge KG, Willinsky RA, Wallace MC. Clinical course of cranial dural arteriovenous fistulas with long-term persistent cortical venous reflux. *Stroke* 2002;**33**(5):1233–6.
6. Awad IA, Little JR, Akarawi WP, Ahl J. Intracranial dural arteriovenous malformations: factors predisposing to an aggressive neurological course. *J Neurosurg* 1990;**72**(6):839–50.
7. Vanlandingham M, Fox B, Hoit D, Elijovich L, Arthur AS. Endovascular treatment of intracranial dural arteriovenous fistulas. *Neurosurgery* 2014;**74**(Suppl 1):S42–9.
8. Roy D, Raymond J. The role of transvenous embolization in the treatment of intracranial dural arteriovenous fistulas. *Neurosurgery* 1997;**40**(6):1133–41. discussion 1141–1134.
9. Hunter W. Further observations upon a particular species of aneurysms. *Med Obstet Soc Phys Lond* 1762;**2**:390.
10. Chitwood W. John and William Hunter on aneurysms. *Arch Surg* 1977;**112**(7):829–36.
11. Perrini P, Nannini T, Di Lorenzo N. Francesco Rizzoli (1809-1880) and the elusive case of Giulia: the description of an "arteriovenous aneurysm passing through the wall of the skull". *Acta Neurochir* 2007;**149**(2):191–6. discussion 196.
12. Dandy WE. Arteriovenous aneurysm of the brain. *Arch Surg* 1928;**17**:190–243.
13. Dandy WE. Carotid-cavernous aneurysms (pulsating exophthalmos). *Zentralbl Neurochir* 1937;**2**:77–113. and 165–206.
14. Cushing H, Bailey P. *Tumors arising from the blood-vessels of the brain: angiomatous malformation and hemagioblastomas*. Springfield, IL: Charles C. Thomas; 1928.
15. Hanford JM, Wheeler JM. Pulsating exophthalmos. *Ann Surg* 1930;**92**(1):8–22.
16. Hamby WB, Gardner WJ. Treatment of pulsating exophthalmos: with report of two cases. *Arch Surg* 1933;**27**(4):676–85.
17. Sundt Jr TM, Piepgras DG. The surgical approach to arteriovenous malformations of the lateral and sigmoid dural sinuses. *J Neurosurg* 1983;**59**(1):32–9.
18. Hugosson R, Bergstrom K. Surgical treatment of dural arteriovenous malformation in the region of the sigmoid sinus. *J Neurol Neurosurg Psychiatry* 1974;**37**(1):97–101.
19. Dawson III RC, Joseph GJ, Owens DS, Barrow DL. Transvenous embolization as the primary therapy for arteriovenous fistulas of the lateral and sigmoid sinuses. *AJNR Am J Neuroradiol* 1998;**19**(3):571–6.
20. Kim DJ, Kim DI, Suh SH, et al. Results of transvenous embolization of cavernous dural arteriovenous fistula: a single-center experience with emphasis on complications and management. *AJNR Am J Neuroradiol* 2006;**27**(10):2078–82.
21. Klisch J, Huppertz HJ, Spetzger U, Hetzel A, Seeger W, Schumacher M. Transvenous treatment of carotid cavernous and dural arteriovenous fistulae: results for 31 patients and review of the literature. *Neurosurgery* 2003;**53**(4):836–56. discussion 856–837.
22. Rivet DJ, Goddard III JK, Rich KM, Derdeyn CP. Percutaneous transvenous embolization of a dural arteriovenous fistula through a mastoid emissary vein. Technical note. *J Neurosurg* 2006;**105**(4):636–9.
23. Gross BA, Albuquerque FC, Moon K, McDougall CG. Evolution of treatment and a detailed analysis of occlusion, recurrence, and clinical outcomes in an endovascular library of 260 dural arteriovenous fistulas. *J Neurosurg* 2017;**126**(6):1884–93.
24. Cognard C, Januel AC, Silva Jr NA, Tall P. Endovascular treatment of intracranial dural arteriovenous fistulas with cortical venous drainage: new management using Onyx. *AJNR Am J Neuroradiol* 2008;**29**(2):235–41.

25. Liebig T, Henkes H, Brew S, Miloslavski E, Kirsch M, Kuhne D. Reconstructive treatment of dural arteriovenous fistulas of the transverse and sigmoid sinus: transvenous angioplasty and stent deployment. *Neuroradiology* 2005;**47**(7):543–51.
26. Borden JA, Wu JK, Shucart WA. A proposed classification for spinal and cranial dural arteriovenous fistulous malformations and implications for treatment. *J Neurosurg* 1995;**82**(2):166–79.
27. Urtasun F, Biondi A, Casaco A, et al. Cerebral dural arteriovenous fistulas: percutaneous transvenous embolization. *Radiology* 1996;**199**(1):209–17.
28. Houdart E, Saint-Maurice JP, Chapot R, et al. Transcranial approach for venous embolization of dural arteriovenous fistulas. *J Neurosurg* 2002;**97**(2):280–6.
29. Lekkhong E, Pongpech S, Ter Brugge K, et al. Transvenous embolization of intracranial dural arteriovenous shunts through occluded venous segments: experience in 51 patients. *AJNR Am J Neuroradiol* 2011;**32**(9):1738–44.
30. Benndorf G, Bender A, Lehmann R, Lanksch W. Transvenous occlusion of dural cavernous sinus fistulas through the thrombosed inferior petrosal sinus: report of four cases and review of the literature. *Surg Neurol* 2000;**54**(1):42–54.
31. Roa JA, Ortega-Gutierrez S, Martinez-Galdamez M, et al. Transcirculation approach for endovascular embolization of intracranial aneurysms, arteriovenous malformations, and dural fistulas: a multicenter study. *World Neurosurg* 2020;**134**:e1015–27.
32. Jittapiromsak P, Ikka L, Benachour N, Spelle L, Moret J. Transvenous balloon-assisted transarterial Onyx embolization of transverse-sigmoid dural arteriovenous malformation. *Neuroradiology* 2013;**55**(3):345–50.
33. Caragine LP, Halbach VV, Dowd CF, Ng PP, Higashida RT. Parallel venous channel as the recipient pouch in transverse/sigmoid sinus dural fistulae. *Neurosurgery* 2003;**53**(6):1261–6. discussion 1266–1267.
34. de Paula Lucas C, Prandini MN, Spelle L, Piotin M, Mounayer C, Moret J. Parallel transverse-sigmoid sinus harboring dural arteriovenous malformation. How to differentiate the pathological and normal sinus in order to treat and preserve patency and function. *Acta Neurochir* 2010;**152**(3):523–7.

# CHAPTER 7

# Carotid-cavernous fistulas

**Jacob F. Baranoski, Felipe C. Albuquerque, and Andrew F. Ducruet**
Department of Neurosurgery, Barrow Neurological Institute, St. Joseph's Hospital and Medical Center, Phoenix, AZ, United States

## Introduction and pertinent anatomy

Carotid-cavernous fistulas (CCFs) are pathologic shunts between the carotid artery and the cavernous sinus. Before discussing CCFs, it is important to recognize that the cavernous sinus is not a structure with a single, capacious lumen but rather comprises numerous thin-walled veins intercalated with neurovascular structures that include the cavernous segment of the internal carotid artery (ICA) and cranial nerves (CN) III, IV, V1, V2, and VI.[1] The cavernous sinus can be divided into four areas—medial, lateral, anteroinferior, and posterosuperior—on the basis of the relative location of the cavernous ICA.[2] Each of these four areas receives blood from specified venous channels. The anterior compartment of the cavernous sinus receives blood from the superior ophthalmic vein (SOV) and inferior ophthalmic vein (IOV). The posterior compartment of the cavernous sinus receives blood from the superior and inferior petrosal veins. The lateral compartment receives blood from the deep middle cerebral vein, the sylvian vein, and the sphenoparietal sinus. In the medial area, the intercavernous sinus and the basilar plexus form two anastomoses between the left and right cavernous sinuses. This comprehension of the anatomy is critical for understanding CCF pathology as well as selecting and executing the appropriate treatment strategy for these challenging lesions.

## CCF classification and etiology

CCFs can be broadly classified on the basis of hemodynamics (high flow vs low flow), etiology (spontaneous vs traumatic), and anatomy.[3] CCFs can be grouped into two anatomical categories, direct and indirect.[4] The Barrow classification further specifies CCFs into four distinct subtypes (A, B, C, and D) on the basis of arterial supply (Table 1).[5] This anatomic definition, and more precisely the Barrow classification system, is preferred because it has pathophysiologic and treatment implications.

*Cerebral Dural Arteriovenous Fistulas*
https://doi.org/10.1016/B978-0-12-819525-3.00001-0

**Table 1** Anatomical classification system for carotid-cavernous fistulas.

| Barrow type | Direct or indirect | Description |
|---|---|---|
| A | Direct | Direct connections between the cavernous ICA and the cavernous sinus |
| B | Indirect | Connection between dural branches of the ICA and the cavernous sinus |
| C | Indirect | Connection between dural branches of the ECA and the cavernous sinus |
| D | Indirect | Connection between dural branches of both the ICA and ECA and the cavernous sinus |

*ECA*, external carotid artery; *ICA*, internal carotid artery.

## Direct CCFs

Direct CCFs (dCCFs), or Barrow Type A CCFs, arise from a defect in the arterial wall of the cavernous ICA that results in a direct, high-flow arteriovenous shunt between the cavernous ICA and cavernous sinus.[5] dCCFs are significantly more common than indirect fistulas, accounting for almost 80% of all CCFs in some series.[5–7] The cause of dCCFs can be (1) posttraumatic (the most common cause) following a closed head injury, with or without a basilar skull fracture, or a penetrating injury; (2) spontaneous, often due to a ruptured cavernous ICA aneurysm[8–10] or genetic conditions that predispose patients to vascular injuries, such as Ehlers-Danlos syndrome[11] or fibromuscular dysplasia[12]; or (3) iatrogenic injury to the cavernous ICA during open microsurgery, transsphenoidal approaches,[13] sinus surgery,[14] or endovascular procedures.[15]

Posttraumatic dCCFs may become symptomatic immediately following the trauma or present in a delayed fashion. In general, the trauma is usually severe and often accompanied by a basilar skull fracture, although this is not always the case.[16] Therefore the exact mechanism that results in a posttraumatic dCCF is unclear, and hypotheses range from a direct tear in the ICA from a bony fracture or shear forces to a sudden elevation in intraluminal pressure within the ICA that results in vessel wall rupture.[3,16] It is important to recognize that posttraumatic dCCFs can be bilateral.[6,17]

## Indirect CCFs

Indirect CCFs are low-flow lesions that are the result of pathologic arteriovenous shunting from meningeal branches of either the ICA or the external carotid artery (ECA) into the cavernous sinus.[4] Indirect CCFs are more appropriately regarded as dural arteriovenous fistulas because these lesions are located in the walls of the cavernous sinus and comprise multiple connections between the cavernous sinus and carotid meningeal branches. On the basis of the exact source of arterial supply, indirect CCFs can be further categorized into Barrow Type B, C, or D: Type B indirect CCFs arise from meningeal branches of the ICA, Type C CCFs arise from meningeal branches of the ECA, and Type D CCFs arise from meningeal branches of both the ICA and ECA (Fig. 1).[5]

Overall, indirect CCFs are significantly less common than dCCFs.[7] The most common arteries identified as the source of the pathologic shunting in indirect CCFs are the internal maxillary, middle meningeal, meningohypophyseal trunk, and capsular arteries.[18]

The exact mechanism that results in the creation of indirect CCFs remains to be fully elucidated. The prevailing theory postulated the formation of microscopic tears in the dural arteries that traverse the cavernous sinus due to microscopic venous thrombosis or increases in venous sinus pressure.[3,5,19] Factors that may increase the risk of these arterial defects and subsequent formation of indirect CCFs include hypertension,[20] pregnancy,[21] diabetes,[20] and genetic conditions that predispose patients to vascular injuries, such as Ehlers-Danlos syndrome[11] or fibromuscular dysplasia.[12] Although the vast majority of CCFs caused by trauma are dCCFs, posttraumatic indirect CCFs have been reported as well.[22] In contrast to dCCFs, indirect CCFs can have contralateral or bilateral feeders and may require bilateral angiography of the ICA and ECA for complete diagnosis and treatment.[9,23]

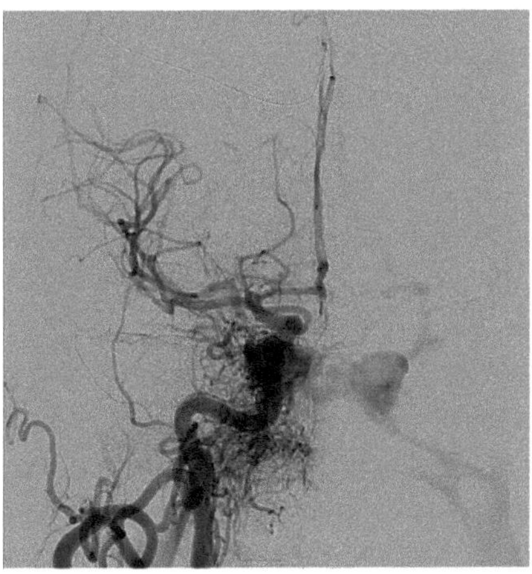

**Fig. 1** Anterior-posterior projection of a right common carotid angiogram showing a complex Barrow Type D indirect CCF. *(Used with permission from Barrow Neurological Institute, Phoenix, Arizona.)*

## Clinical presentation

The clinical features associated with CCFs are the sequelae of the resultant hemodynamic dysfunction within the cavernous sinus. The precise symptoms are based primarily on the rate of flow of the shunt and the location or pattern of venous drainage.[24]

dCCFs are high-flow lesions that typically present in an acute fashion, which is consistent with their mechanism of formation following severe trauma or aneurysmal rupture, and are associated with a very low likelihood of spontaneous resolution.[3,6,8,9] The acute onset and rapid progression associated with these lesions often necessitate urgent treatment. dCCFs often present with the characteristic features of proptosis, chemosis, orbital bruits, visual deficits, and headache that are the result of orbital vein arterialization.[3,9,25] Cranial nerve deficits resulting in ophthalmoplegia and diplopia as well as other cranial nerve deficits, including vision loss, are also frequently present; less commonly, dCCFs can present with an acute intracerebral hemorrhage.[3,26]

In contrast to dCCFs, indirect CCFs are comparatively low-flow lesions that can have a more insidious, gradual onset with a constellation of symptoms that may relapse and remit.[27] These variable symptoms can lead to delays in diagnosis and subsequent treatment. For indirect CCFs, the pattern of venous drainage is directly related to the clinical presentation.[24,27] Indirect CCFs that drain posteriorly into the superior or inferior petrosal sinuses are typically less symptomatic and may be clinically silent. If posteriorly draining indirect CCFs are symptomatic, they can present with face pain, facial paralysis, or ophthalmoplegia.[24,27]

Anteriorly draining indirect CCFs are more common and typically present with ocular and orbital symptoms, such as chemosis, conjunctival injection, and proptosis.[24,27] These symptoms can resemble those of dCCFs but are often milder and less abrupt in onset. Cranial nerve deficits resulting in diplopia, vision loss, and ophthalmoplegia may also be present.[24,27] Indirect CCFs that are associated with retrograde cortical venous reflux often present with neurologic symptoms referable to the exact cortical or subcortical location of venous congestion, hemorrhage, or infarction. Contributing to the potential delay in diagnosis, the pattern of venous drainage of indirect CCFs has been demonstrated to change with the development and resolution of cavernous sinus thromboses, which results in a varying pattern of symptoms.[3,27]

## Radiologic diagnosis

A high index of suspicion is essential for early recognition and diagnosis of these lesions, because symptom onset may be gradual and insidious. For patients who present with the aforementioned symptoms, noninvasive radiographic imaging, including computed tomography, magnetic resonance imaging, and/or computed tomography/magnetic resonance angiography, can be helpful in establishing the diagnosis of a CCF. Evidence of cavernous sinus enlargement, SOV dilation, or engorgement of cortical or meningeal vessels can hint at the presence of an underlying CCF. Furthermore, a basal skull fracture or cavernous ICA aneurysm should raise the index of suspicion for an associated dCCF.[3, 19, 25–27] However, it is important to note that the absence of abnormal findings on these noninvasive imaging modalities does not exclude the diagnosis of a CCF, and the gold standard for diagnosis of these lesions remains the catheter cerebral angiogram.

When a CCF is suspected, a cerebral angiogram that includes injection of bilateral ICAs, bilateral ECAs, and at least the dominant vertebral artery should be performed.[28] Bilateral evaluation is critical because CCFs can often involve feeding vessels from the contralateral ICA. Catheter angiography is necessary to assess the angioarchitecture of the lesion, feeding arteries, and venous drainage pattern and is essential for treatment planning.

Two important angiographic maneuvers that can be useful in delineating the angioarchitecture of CCFs, particularly for very high-flow lesions, are the Mehringer-Hieshima and Huber maneuvers. The Mehringer-Hieshima maneuver is used to limit the amount of contrast entering the high-flow shunt to better elucidate the venous drainage pattern and is performed by slowly injecting the ipsilateral ICA while applying gentle manual compression to the artery to limit the flow of contrast into the ICA.[29] The Huber maneuver refers to compression applied to the ipsilateral carotid artery during vertebral artery injection to better demonstrate shunting from the posterior circulation vessels.[30]

## CCF treatment

The ultimate goal of treatment for CCFs is to completely occlude or obliterate the fistula while preserving or restoring normal arterial flow through the ICA. As with most pathologic conditions, CCF treatment has evolved drastically over the years due to significant improvement in medical technologies and techniques. Historically, CCFs were treated with intermittent compression of the ipsilateral carotid artery or surgical trapping or ligation of the ICA.[31, 32] As endovascular techniques became available, CCFs started to be treated with detachable silicone balloons designed to occlude the artery and prevent shunting.[25] This was a popular and relatively effective technique for treating CCFs and was used until the elimination of these devices from the North American market in 2003. With the advent of an ever-expanding armamentarium of new endovascular devices and the development and refinement of endovascular techniques, endovascular intervention has become the primary and usually definitive treatment for these lesions. Compared with the interventions mentioned above, these newer techniques and technologies enable surgeons and interventionalists to achieve the primary objective in treating CCFs, complete occlusion of the shunt and preservation of normal ICA flow. The various strategies used to treat CCFs account for differences in the size and anatomic location of the pathologic shunt as well as variations in the angioarchitecture, including arterial inflow dynamics and patterns of venous outflow. Meticulous inspection of the anatomy using a pretreatment angiogram is essential, and a tailored, patient-specific approach should be taken for each CCF. In the following subsections, we discuss the evolving array of endovascular techniques performed to safely and effectively treat CCFs. We also discuss the potential roles for conservative therapy, microsurgery, and radiosurgery for these lesions.

## Conservative management

Low-flow indirect CCFs that present with mild symptoms and no indications of the need for emergent intervention may be successfully managed by external manual compression of the ipsilateral cervical carotid artery and jugular vein.[31, 33] Compression is applied throughout the day using the contralateral hand for 4–6 weeks. This technique decreases the rate and extent of flow through the pathologic shunt and promotes spontaneous thrombosis of the fistula.[31, 33] Again, this technique is reserved for cases of low-flow, indirect CCFs without severe or acute symptoms. This technique is unwarranted and ineffective in treating high-flow CCFs or dCCFs. It is worth noting that the incidence of spontaneous resolution of low-flow indirect CCFs without treatment has been reported to be as high as 50% in some series.[27, 34] This technique of intermittent compression therefore likely capitalizes on and promotes the natural history of this particular subset of these lesions.

A conservative treatment strategy must be used in conjunction with close clinical follow-up, including serial neurologic and ophthalmologic examinations. Progressive visual decline, development or worsening of cranial nerve deficits, or an elevation of intraocular pressure should prompt urgent definitive intervention. Likewise, patients who do not improve clinically following a completed course of conservative management or have documented patency of the fistula on follow-up imaging should be referred for definitive treatment.

## Endovascular treatment

Endovascular innovations continue to drive the evolution of CCF treatment. Endovascular treatment strategies now include both transarterial and transvenous access routes, with embolization using an array of detachable coils and liquid embolic agents. Furthermore, adjunctive techniques, including new stent and flow-diversion technologies, are being used to help preserve the patency of ICA and decrease the rate of flow across the pathologic shunt to facilitate transvenous treatment delivery. Selection of the specific endovascular approach is based on the angiographic features of each individual fistula. Decision-making begins with a detailed angiographic analysis of the fistula, including study of the feeding arterial vessels and venous drainage, which informs treatment and approach selection. Because of improvement in these technologies and techniques, endovascular intervention has become the primary and usually definitive treatment for CCFs, with safe and effective endovascular strategies now available for even the most complex lesions. Overall, long-term endovascular cure rates for both indirect CCFs and dCCFs in recent series exceed 90%, and both treatments are associated with low complication profiles and favorable clinical outcomes.[3, 6, 8, 9, 26, 27, 35–37]

## Endovascular techniques for dCCFs

Improvements in endovascular therapies in recent decades have led to a significant evolution in the treatment algorithm for dCCFs.[8,9,38,39] Transarterial approaches have historically represented the primary treatment for dCCFs. In the United States, detachable silicone balloons were commonly and effectively used to treat dCCFs.[25] After 2003, however, these devices were withdrawn from the North American market, and neurosurgeons and interventionalists began to use detachable coils to treat CCFs.[40] In cases where the precise location of the fistula can be clearly identified, such as shunts related to a ruptured cavernous ICA aneurysm with a well-defined neck, primary coil embolization may be effective in achieving fistula closure.[3, 9] However, in cases in which the exact site and size of the pathologic shunt are more difficult to elucidate, balloon remodeling and/or stent placement may be required to preserve ICA patency while achieving obliteration of the fistula.[9, 35] More recently, transarterial approaches are being used in combination with transvenous approaches to achieve a safer and more efficacious treatment for dCCFs.[3, 9, 26, 35] Transvenous embolization can assist in achieving complete obliteration of the fistula, whereas concurrent transarterial access can afford ICA protection with either a balloon or a stent. The use of liquid embolic agents, including *n*-butyl-2-cyanoacrylate (NBCA) and ethylene-vinyl alcohol copolymer (Onyx) has also been reported,[36, 41, 42] and a recent study has demonstrated improved radiographic outcomes associated with dCCFs when a liquid embolic is used in conjunction with detachable coils compared with coils alone.[43] Liquid embolic agents can penetrate the small channels of the cavernous sinus more effectively than can detachable coils and therefore result in more complete obliteration of the fistula. However, the use of liquid embolic agents is also associated with a higher risk of off-target embolization, and therefore care must be taken to minimize this risk. Concurrent transarterial access can afford ICA protection and decrease the risk of off-target embolization by temporarily occluding the shunt. The initial placement of several detachable coils within the cavernous sinus to further decrease the rate and amount of flow across the fistula can also serve as a framework to help ensnare the liquid embolic agents.[9, 26, 35] Our group and others have reported favorable outcomes utilizing this combined transarterial/transvenous approach.[3, 9, 26, 35]

The development of flow-diverting devices, including the pipeline embolization device (PED), has provided another option in the armamentarium for treating dCCFs.[8, 15, 38, 39, 44–48] The flow-diverting properties of the PED and other flow-diverter stents lend themselves well to treating these high-flow shunts. By promoting laminar flow through the parent artery, these devices can divert blood away from the fistula while simultaneously serving as a scaffold that provides a physical and hemodynamic barrier to help prevent herniation or migration of transvenously or transarterially deployed embolic material. We recently published a series of dCCFs for which we utilized the PED as the upfront, primary treatment modality to achieve safe and effective obliteration of the fistula.[8] We found that the PED independently decreased the shunting across the fistula, resulted in stasis within the cavernous sinus, and promoted normalization of distal arterial flow (Fig. 2A–D). Furthermore, in a subset of patients who underwent adjunctive transvenous treatment, the PED was the only arterial protection device used, and an arterial balloon was not required to prevent off-target deposition of transvenously delivered liquid embolic agents. Other groups have also reported use of the PED as a salvage technique for intraprocedural iatrogenic injury or as an adjunctive treatment strategy for residual or recurrent fistulas after unsuccessful primary treatment of dCCFs.[39, 44–49] Given these results, we have adopted a strategy of upfront use of the PED for the treatment of dCCFs. However, caution

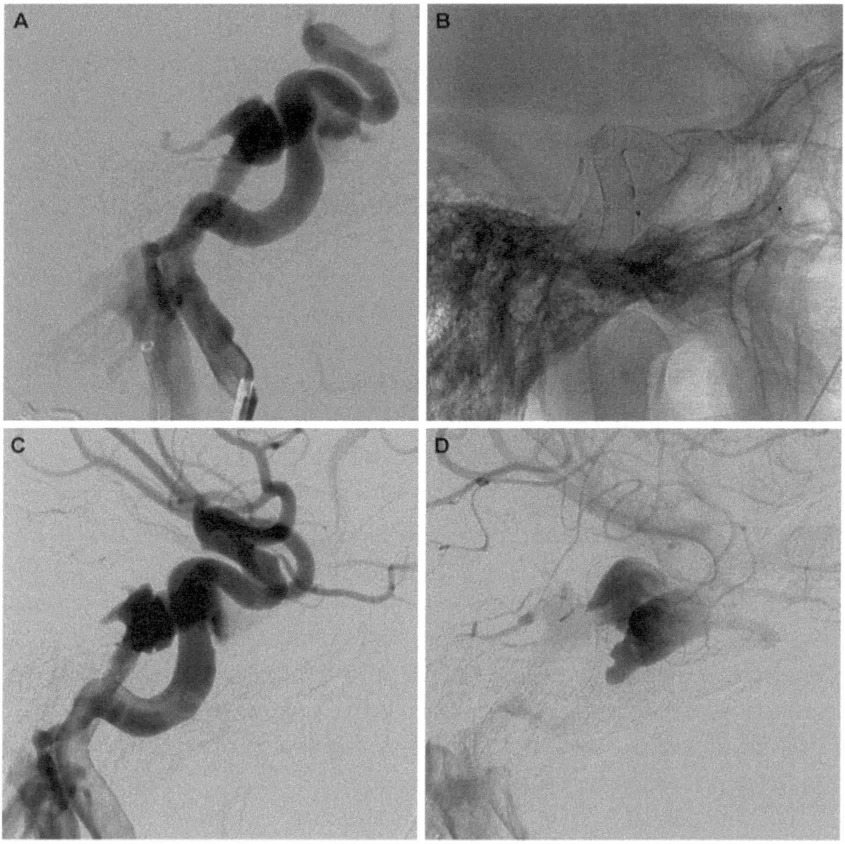


**Fig. 2** (A) Lateral projection of a right ICA angiogram showing a right dCCF. (B) Unsubtracted lateral projection showing placement of a PED across a fistulous site. (C) Lateral projection of a right ICA angiogram showing decreased shunting across the dCCF and increased ipsilateral normal distal carotid flow after PED deployment. (D) Lateral projection of a right ICA angiogram showing stagnation of flow within the cavernous sinus after PED deployment. *(Used with permission from Barrow Neurological Institute, Phoenix, Arizona.)*

must be exercised in the setting of multisystem traumatic injuries because of the current need for dual antiplatelet therapy with the PED. Improvement in flow-diverting devices should limit the requirement for dual antiplatelet therapy and further expand the utility of flow-diverting devices for treatment of these lesions. As a salvage technique for the increasingly rare cases in which the fistula cannot be occluded while ICA patency is maintained, ICA sacrifice using coil embolization can be performed if the patient has passed a balloon-test occlusion or in emergent situations.[3, 9, 26, 35]

## Endovascular techniques for indirect CCFs

In contrast to dCCFs, for which a transarterial approach remains a mainstay of the intervention, indirect CCFs often cannot be treated effectively with a purely transarterial strategy. Unlike dCCFs, in which there is typically a single fistulous channel between the ICA and cavernous sinus, indirect CCFs involve multiple small arterial connections from the ICA, ECA, or both. Consequently a purely transarterial treatment is both less efficacious due to incomplete obliteration of all the feeding vessels and more hazardous due to an increased risk of off-target embolization.[3, 9, 26, 27]

Therefore the primary treatment strategy for indirect CCFs involves a transvenous or combined transvenous-transarterial approach. The goal of intervention for indirect CCFs, similar to that for dCCFs, is obliteration of the fistula, preservation of ICA patency, and limitation of off-target embolization. This is achieved via obliteration of the cavernous sinus and eradication of the shunts. Transvenous access to the cavernous sinus has been conventionally achieved via inferior petrosal sinus (IPS).[18, 27, 40] Even in cases where IPS drainage is not apparent or the IPS appears to be thrombosed, it is often still possible, thanks to improvements in microcatheter technology and techniques, to navigate a microwire and catheter through the stenotic or thrombosed IPS to reach the cavernous sinus. Once there, embolization can be achieved using a combination of coil and/or liquid embolic agents (further discussed later).[3, 6, 9, 18, 26, 27, 35, 36, 40, 43] As with dCCFs, safe and effective transvenous embolization of indirect CCFs can be facilitated by using a concurrent transarterial approach to provide ICA protection via either a temporarily inflated balloon or stent deployment.[6, 9, 35] Regardless of whether adjunctive transarterial techniques are used, transarterial access should always be achieved to perform control angiography and assess whether the fistula has been obliterated. As flow-diverter technology continues to improve, it likely will be useful in reconstructing laminar flow through the carotid artery while shunting blood away from these low-flow fistulas to further promote obliteration of these lesions.[8, 9, 38]

For cases where the IPS cannot be accessed, alternative transvenous routes to the cavernous sinus can be accessed via femoral puncture. These options include the superior petrosal sinus, pterygoid plexus, basilar plexuses, or even contralateral structures.[3, 6, 9, 23, 26, 27, 35, 50] Recent improvements in catheter technology as well as in endovascular technique have allowed for additional access and CCF treatment via the orbital transvenous routes. These approaches utilize the SOV, IOV, or medial ophthalmic vein as a corridor to the cavernous sinus and can be accessed via the transfemoral route through the facial vein (Fig. 3A and B).[9, 35, 37] The SOV can be accessed via direct percutaneous puncture under fluoroscopic guidance.[9, 35, 37, 51] Additional options that can be utilized if access is difficult include cutdown to access the SOV, orbitotomy or craniotomy to access the SOV or cortical vein, and direct transorbital puncture into the cavernous sinus.[3,6,9,26,27,35,37,50] These transorbital venous approaches have been used for some time but were historically less preferred because of concern for increased risk of complications compared with the more traditional transfemoral approaches through the IPS.[52] However, a recent meta-analysis found all the aforementioned

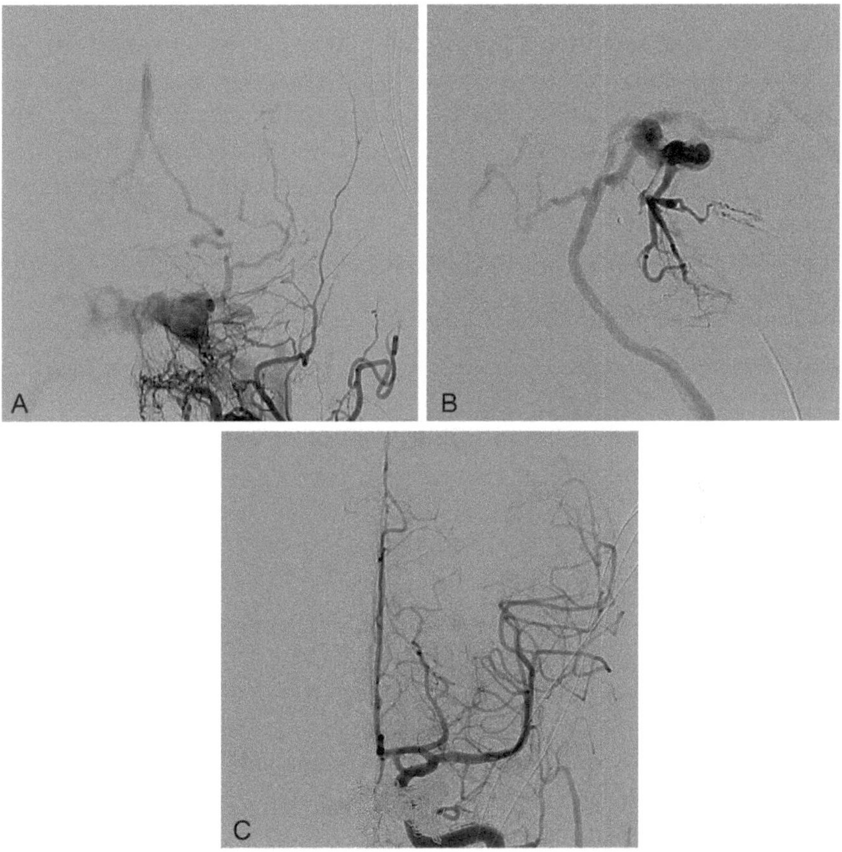


**Fig. 3** (A) Anterior-posterior projection of a left common carotid angiogram showing a left indirect CCF. (B) Anterior-posterior projection of a left SOV injection showing placement of microcatheter in the left SOV with opacification of the left cavernous sinus. (C) Anterior-posterior projection of a left common carotid angiogram showing obliteration of the left indirect CCF following transvenous coil embolization via the left SOV. *(Used with permission from Barrow Neurological Institute, Phoenix, Arizona.)*

transorbital access routes to be safe and effective in treating CCFs.[37] We advocate for their use in situations where standard transfemoral access is not feasible or the angioarchitecture of the fistula necessitates access via the SOV, but this cannot be achieved via the facial vein. For cases that cannot be accessed transfemorally, we favor percutaneous orbital puncture under fluoroscopic guidance.

With a microcatheter navigated to the cavernous sinus, transvenous embolization can be performed using detachable coils and/or liquid embolic agents (Fig. 3C). Liquid embolic agents can penetrate the small channels of the cavernous sinus more effectively than can detachable coils and therefore can achieve a more complete obliteration of the fistula.[3, 6, 9, 26, 35, 36, 50] A recent study demonstrated that, when a liquid embolic is used in conjunction with detachable coils, treatment outcomes are superior compared with when CCFs are treated with coils alone.[43] However, use of liquid embolic agents is also associated with a higher risk of off-target embolization.[3,6,9,26,35,36,50] Concurrent transarterial access can afford ICA protection and decrease this risk while also allowing for the performance of frequent control angiograms to assess treatment efficacy and extent of off-target embolic agent deposition.

As with selecting the appropriate approach when treating a CCF, determining which embolic agent to use should account for the specific angioarchitecture of the lesion and each fistula should receive a tailored, patient-specific treatment. Currently, Onyx and NBCA are the two most commonly used liquid embolic agents. Particle embolization is no longer recommended because of a high rate of recanalization.[9, 50] Onyx permits slow injection and affords the opportunity to perform multiple control angiograms to assess treatment progress and off-target accumulation. The embolic properties of Onyx with its "lava-like" deposition can be advantageous in penetrating the trabeculated channels of the cavernous sinus. In contrast to Onyx, NBCA has rapid polymerization that does not permit slow injection and limits real-time assessment of embolization progress. In general, we prefer to use Onyx rather than NBCA for CCF treatment. Regardless of the liquid embolic agent being used, we strongly advocate for the use of detachable-tip catheters to prevent trapping of the catheter in the embolic cast.

## Microsurgical treatment

Despite significant improvement in endovascular techniques, definitive treatment is not always achieved. In these increasingly rare cases, alternative treatment modalities, including microsurgery and radiosurgery, should be considered. Although microsurgical treatment for CCF has become increasingly uncommon, it remains a viable salvage option.[3, 9, 26, 32, 35, 53] Surgical intervention may involve suturing, clipping, or trapping of the fistula; direct packing of the cavernous sinus; ligating the ICA; or a combination of these maneuvers. Day et al.[32] and Tu et al.[53] each reported successful microsurgical treatment of CCFs in which endovascular treatment failed while maintaining ICA patency in approximately 75% of patients. Extracranial to intracranial bypass surgery may also be necessary for emergent cases where ICA sacrifice is required but the patient has a failed balloon occlusion test or the test could not be performed because of the emergent nature of the case.

### Radiosurgery

As with other types of complex dural arteriovenous fistula, radiosurgery may be used to treat patients with indirect CCF. Radiosurgery has been used to treat patients with low-flow fistulas with relatively good success rates that approach a 90% rate of obliteration and an 85% rate of symptomatic improvement.[3, 9, 26, 35, 54] Radiosurgery can be used as stand-alone treatment for patients who are poor candidates for endovascular procedures, in conjunction with endovascular intervention, or as a salvage technique for patients after failed or incomplete endovascular treatment. Radiosurgery should not be performed as the primary modality on dCCFs or high-flow lesions and should not be performed for cases with acute and severe symptom onset or progression because the latency between treatment and clinical resolution can be delayed by almost 2 years.

## Outcomes and prognosis

Overall, thanks primarily to improvements in endovascular technologies and techniques, successful treatment of even the most complex CCFs can now be achieved. In general, recent series have reported cure rates of greater than 90% for both indirect CCFs and dCCFs.[3, 8, 9, 18, 26, 27, 35] Following successful treatment, symptoms such as chemosis and proptosis generally resolve within hours to days, and cranial nerve deficits typically resolve over the course of several weeks. The extent of vision recovery is largely dependent on the pathogenesis, severity, and duration of the preintervention deficit. Chances of recovery decrease if the SOV is already thrombosed or the central retinal vein has already been affected at the time of the diagnosis. Recurrence of CCFs following complete angiographic obliteration is rare.

## Conclusions

CCFs, although historically difficult to treat, are now routinely managed and are associated with low rates of morbidity and mortality. Endovascular intervention with a goal of complete fistula occlusion and preservation of normal blood flow through the internal carotid artery has emerged as the primary treatment modality. Complete fistula obliteration with associated symptom improvement or resolution can be achieved in most cases after appropriate therapy. Meticulous inspection of the anatomy and angioarchitecture is essential, and a tailored, patient-specific approach should be taken for each CCF.

## Acknowledgments

The authors thank the staff of Neuroscience Publications at Barrow Neurological Institute for assistance with manuscript preparation.

## Disclosures

None.

## Financial support

None.

## References

1. Parkinson D. Lateral sellar compartment O.T. (cavernous sinus): history, anatomy, terminology. *Anat Rec* 1998;**251**(4):486–90 [Epub 1998/08/26].
2. Harris FS, Rhoton AL. Anatomy of the cavernous sinus. A microsurgical study. *J Neurosurg* 1976;**45**(2):169–80 [Epub 1976/08/01].
3. Ellis JA, Goldstein H, Connolly Jr ES, Meyers PM. Carotid-cavernous fistulas. *Neurosurg Focus* 2012;**32**(5):E9 [Epub 2012/04/28].
4. Peeters FL, Kroger R. Dural and direct cavernous sinus fistulas. *AJR Am J Roentgenol* 1979;**132**(4):599–606 [Epub 1979/04/01].
5. Barrow DL, Spector RH, Braun IF, Landman JA, Tindall SC, Tindall GT. Classification and treatment of spontaneous carotid-cavernous sinus fistulas. *J Neurosurg* 1985;**62**(2):248–56 [Epub 1985/02/01].
6. Korkmazer B, Kocak B, Tureci E, Islak C, Kocer N, Kizilkilic O. Endovascular treatment of carotid cavernous sinus fistula: a systematic review. *World J Radiol* 2013;**5**(4):143–55 [Epub 2013/05/15].
7. Debrun GM, Vinuela F, Fox AJ, Davis KR, Ahn HS. Indications for treatment and classification of 132 carotid-cavernous fistulas. *Neurosurgery* 1988;**22**(2):285–9 [Epub 1988/02/01].

8. Baranoski JF, Ducruet AF, Przbylowski CJ, Almefty RO, Ding D, Catapano JS, et al. Flow diverters as a scaffold for treating direct carotid cavernous fistulas. *J Neurointerv Surg* 2019;**11**(11):1129–34 [Epub 2019/07/04].
9. Ducruet AF, Albuquerque FC, Crowley RW, McDougall CG. The evolution of endovascular treatment of carotid cavernous fistulas: a single-center experience. *World Neurosurg* 2013;**80**(5):538–48 [Epub 2013/02/14].
10. Linskey ME, Sekhar LN, Hirsch Jr W, Yonas H, Horton JA. Aneurysms of the intracavernous carotid artery: clinical presentation, radiographic features, and pathogenesis. *Neurosurgery* 1990;**26**(1):71–9 [Epub 1990/01/01].
11. Farley MK, Clark RD, Fallor MK, Geggel HS, Heckenlively JR. Spontaneous carotid-cavernous fistula and the Ehlers-Danlos syndromes. *Ophthalmology* 1983;**90**(11):1337–42 [Epub 1983/11/01].
12. Hirai T, Korogi Y, Goto K, Ogata N, Sakamoto Y, Takahashi M. Carotid-cavernous sinus fistula and aneurysmal rupture associated with fibromuscular dysplasia. A case report. *Acta Radiol* 1996;**37**(1):49–51 [Epub 1996/01/01].
13. Pigott TJ, Holland IM, Punt JA. Carotico-cavernous fistula after trans-sphenoidal hypophysectomy. *Br J Neurosurg* 1989;**3**(5):613–6 [Epub 1989/01/01].
14. Pedersen RA, Troost BT, Schramm VL. Carotid-cavernous sinus fistula after external ethmoid-sphenoid surgery. Clinical course and management. *Arch Otolaryngol* 1981;**107**(5):307–9 [Epub 1981/05/01].
15. Alan N, Nwachuku E, Jovin TJ, Jankowitz BT, Jadhav AP, Ducruet AF. Management of iatrogenic direct carotid cavernous fistula occurring during endovascular treatment of stroke. *World Neurosurg* 2017;**100**:710.e15–20 [Epub 2017/02/12].
16. Helmke K, Kruger O, Laas R. The direct carotid cavernous fistula: a clinical, pathoanatomical, and physical study. *Acta Neurochir* 1994;**127**(1–2):1–5 [Epub 1994/01/01].
17. Liang W, Xiaofeng Y, Weiguo L, Wusi Q, Gang S, Xuesheng Z. Traumatic carotid cavernous fistula accompanying basilar skull fracture: a study on the incidence of traumatic carotid cavernous fistula in the patients with basilar skull fracture and the prognostic analysis about traumatic carotid cavernous fistula. *J Trauma* 2007;**63**(5):1014–20 [discussion 20. Epub 2007/11/13].
18. Meyers PM, Halbach VV, Dowd CF, Lempert TE, Malek AM, Phatouros CC, et al. Dural carotid cavernous fistula: definitive endovascular management and long-term follow-up. *Am J Ophthalmol* 2002;**134**(1):85–92 [Epub 2002/07/04].
19. Ringer AJ, Salud L, Tomsick TA. Carotid cavernous fistulas: anatomy, classification, and treatment. *Neurosurg Clin N Am* 2005;**16**(2):279–95. viii. [Epub 2005/02/08].
20. Oishi A, Miyamoto K, Yoshimura N. Etiology of carotid cavernous fistula in Japanese. *Jpn J Ophthalmol* 2009;**53**(1):40–3 [Epub 2009/02/03].
21. Lin TK, Chang CN, Wai YY. Spontaneous intracerebral hematoma from occult carotid-cavernous fistula during pregnancy and puerperium. Case report. *J Neurosurg* 1992;**76**(4):714–7 [Epub 1992/04/01].
22. Luo CB, Teng MM, Chang FC, Chang CY. Traumatic indirect carotid cavernous fistulas: angioarchitectures and results of transarterial embolization by liquid adhesives in 11 patients. *Surg Neurol* 2009;**71**(2):216–22 [Epub 2008/02/23].
23. Dabus G, Batjer HH, Hurley MC, Nimmagadda A, Russell EJ. Endovascular treatment of a bilateral dural carotid-cavernous fistula using an unusual unilateral approach through the basilar plexus. *World Neurosurg* 2012;**77**(1):201.e5–8 [Epub 2011/12/02].
24. Suh DC, Lee JH, Kim SJ, Chung SJ, Choi CG, Kim HJ, et al. New concept in cavernous sinus dural arteriovenous fistula: correlation with presenting symptom and venous drainage patterns. *Stroke* 2005;**36**(6):1134–9 [Epub 2005/05/14].
25. Lewis AI, Tomsick TA, Tew Jr JM. Management of 100 consecutive direct carotid-cavernous fistulas: results of treatment with detachable balloons. *Neurosurgery* 1995;**36**(2):239–44 [discussion 44–5. Epub 1995/02/01].

26. Zanaty M, Chalouhi N, Tjoumakaris SI, Hasan D, Rosenwasser RH, Jabbour P. Endovascular treatment of carotid-cavernous fistulas. *Neurosurg Clin N Am* 2014;**25**(3):551–63 [Epub 2014/07/06].
27. Miller NR. Dural carotid-cavernous fistulas: epidemiology, clinical presentation, and management. *Neurosurg Clin N Am* 2012;**23**(1):179–92 [Epub 2011/11/24].
28. Debrun GM. Angiographic workup of a carotid cavernous sinus fistula (CCF) or what information does the interventionalist need for treatment? *Surg Neurol* 1995;**44**(1):75–9 [Epub 1995/07/01].
29. Mehringer CM, Hieshima GB, Grinnell VS, Tsai F, Pribram HF. Improved localization of carotid cavernous fistula during angiography. *AJNR Am J Neuroradiol* 1982;**3**(1):82–4 [Epub 1982/01/01].
30. Huber P. A technical contribution of the exact angiographic localization of carotid cavernous fistulas. *Neuroradiology* 1976;**10**(5):239–41 [Epub 1976/01/01].
31. Higashida RT, Hieshima GB, Halbach VV, Bentson JR, Goto K. Closure of carotid cavernous sinus fistulae by external compression of the carotid artery and jugular vein. *Acta Radiol Suppl* 1986;**369**:580–3 [Epub 1986/01/01].
32. Day JD, Fukushima T. Direct microsurgery of dural arteriovenous malformation type carotid-cavernous sinus fistulas: indications, technique, and results. *Neurosurgery* 1997;**41**(5):1119–24 [discussion 24–6. Epub 1997/11/15].
33. Kai Y, Hamada J, Morioka M, Yano S, Kuratsu J. Treatment of cavernous sinus dural arteriovenous fistulae by external manual carotid compression. *Neurosurgery* 2007;**60**(2):253–7 [discussion 7–8].
34. Bujak M, Margolin E, Thompson A, Trobe JD. Spontaneous resolution of two dural carotid-cavernous fistulas presenting with optic neuropathy and marked congestive ophthalmopathy. *J Neuroophthalmol* 2010;**30**(3):222–7.
35. Sur S, Menaker SA, Alvarez C, Chen S, Shah SS, Peterson EC, et al. Multimodal management of carotid-cavernous fistulas. *World Neurosurg* 2020;**133**:e796–803.
36. Barber SM, Rangel-Castilla L, Zhang YJ, Klucznik R, Diaz O. Mid- and long-term outcomes of carotid-cavernous fistula endovascular management with Onyx and n-BCA: experience of a single tertiary center. *J Neurointerv Surg* 2015;**7**(10):762–9.
37. Phan K, Xu J, Leung V, Teng I, Sheik-Ali S, Maharaj M, et al. Orbital approaches for treatment of carotid cavernous fistulas: a systematic review. *World Neurosurg* 2016;**96**:243–51.
38. Patel PD, Chalouhi N, Atallah E, Tjoumakaris S, Hasan D, Zarzour H, et al. Off-label uses of the pipeline embolization device: a review of the literature. *Neurosurg Focus* 2017;**42**(6):E4.
39. Wendl CM, Henkes H, Martinez Moreno R, Ganslandt O, Bazner H, Aguilar Perez M. Direct carotid cavernous sinus fistulae: vessel reconstruction using flow-diverting implants. *Clin Neuroradiol* 2017;**27**(4):493–501.
40. Bink A, Goller K, Luchtenberg M, Neumann-Haefelin T, Dutzmann S, Zanella F, et al. Long-term outcome after coil embolization of cavernous sinus arteriovenous fistulas. *AJNR Am J Neuroradiol* 2010;**31**(7):1216–21.
41. Yu Y, Li Q, Huang Q, Zhang Y, Fang Y, Xu Y, et al. Embolization of direct carotid cavernous fistula with Onyx and coils under transarterial balloon protection. *Cardiovasc Intervent Radiol* 2014;**37**(3):679–85.
42. Yu Y, Huang Q, Xu Y, Hong B, Zhao W, Deng B, et al. Use of onyx for transarterial balloon-assisted embolization of traumatic carotid cavernous fistulas: a report of 23 cases. *AJNR Am J Neuroradiol* 2012;**33**(7):1305–9.
43. de Castro-Afonso LH, Trivelato FP, Rezende MT, Ulhoa AC, Nakiri GS, Monsignore LM, et al. Transvenous embolization of dural carotid cavernous fistulas: the role of liquid embolic agents in association with coils on patient outcomes. *J Neurointerv Surg* 2018;**10**(5):461–2.

44. Amuluru K, Al-Mufti F, Gandhi CD, Prestigiacomo CJ, Singh IP. Direct carotid-cavernous fistula: a complication of, and treatment with, flow diversion. *Interv Neuroradiol* 2016;**22**(5):569–76.
45. Nadarajah M, Power M, Barry B, Wenderoth J. Treatment of a traumatic carotid-cavernous fistula by the sole use of a flow diverting stent. *J Neurointerv Surg* 2012;**4**(3):e1.
46. Nossek E, Zumofen D, Nelson E, Raz E, Potts MB, Desousa KG, et al. Use of pipeline embolization devices for treatment of a direct carotid-cavernous fistula. *Acta Neurochir* 2015;**157**(7):1125–9 [discussion 30].
47. Ogilvy CS, Motiei-Langroudi R, Ghorbani M, Griessenauer CJ, Alturki AY, Thomas AJ. Flow diverters as useful adjunct to traditional endovascular techniques in treatment of direct carotid-cavernous fistulas. *World Neurosurg* 2017;**105**:812–7.
48. Pradeep N, Nottingham R, Kam A, Gandhi D, Razack N. Treatment of post-traumatic carotid-cavernous fistulas using pipeline embolization device assistance. *J Neurointerv Surg* 2016;**8**(10):e40.
49. Iancu D, Lum C, Ahmed ME, Glikstein R, Dos Santos MP, Lesiuk H, et al. Flow diversion in the treatment of carotid injury and carotid-cavernous fistula after transsphenoidal surgery. *Interv Neuroradiol* 2015;**21**(3):346–50.
50. Saraf R, Shrivastava M, Siddhartha W, Limaye U. Evolution of endovascular management of intracranial dural arteriovenous fistulas: single center experience. *Neurol India* 2010;**58**(1):62–8.
51. Dashti SR, Fiorella D, Spetzler RF, Albuquerque FC, McDougall CG. Transorbital endovascular embolization of dural carotid-cavernous fistula: access to cavernous sinus through direct puncture: case examples and technical report. *Neurosurgery* 2011;**68**(1 Suppl. Operative):75–83 [discussion].
52. Uflacker R, Lima S, Ribas GC, Piske RL. Carotid-cavernous fistulas: embolization through the superior ophthalmic vein approach. *Radiology* 1986;**159**(1):175–9.
53. Tu YK, Liu HM, Hu SC. Direct surgery of carotid cavernous fistulae and dural arteriovenous malformations of the cavernous sinus. *Neurosurgery* 1997;**41**(4):798–805 [discussion 6].
54. Pan HC, Sun MH, Sheehan J, Sheu ML, Chen CC, Lee HT, et al. Radiosurgery for dural carotid-cavernous sinus fistulas: Gamma Knife compared with XKnife radiosurgery. *J Neurosurg* 2010;**113**(Suppl):9–20.

# CHAPTER 8

# Surgical management of cerebral dural arteriovenous fistulas

**Tyler S. Cole and Michael T. Lawton**
Department of Neurosurgery, Barrow Neurological Institute, St. Joseph's Hospital and Medical Center, Phoenix, AZ, United States

## Introduction

Cranial vascular lesions present unique anatomic challenges, and dural arteriovenous fistulas (dAVFs) are no exception. Like arteriovenous malformations (AVMs), the core feature of their pathology is abnormal arterial to venous communication leading to shunting (Fig. 1). dAVFs account for approximately 10% of all intracranial vascular malformations.[1, 2] Fed by meningeal arteries, they have a comparably higher rupture risk than AVMs. Understanding the blood flow within these lesions and the different natural histories related to these flow patterns have led clinicians to develop classification systems to guide treatment strategies, discussed in other portions of this book. The clinical manifestations of dAVFs arise from venous hypertension. Congenital dAVFs occur in the pediatric population, though the majority of adult and pediatric dAVFs are acquired.[3–6] Associations with traumatic brain injury, previous surgery, and venous thrombosis of the cerebral sinuses have been established.[4, 6]

The treatment strategies for cerebral dAVFs increasingly incorporate endovascular approaches as minimally invasive techniques have evolved to better target and embolize these lesions. In this chapter, we discuss where open surgical treatment remains an integral part of the management strategy, either for the primary treatment of dAVFs or to supplement endovascular approaches. Fig. 2 provides our general recommendations and broad treatment goals based on dAVF characteristics.

dAVFs such as carotid-cavernous, transverse-sigmoid sinus, superior sagittal sinus, and infratentorial dAVFs are often manageable with endovascular therapy. However, several types of dAVFs may have anatomic restrictions that favor open surgical treatment. If selective catheterization of an anatomic location is challenging or if refluxing embolisate from endovascular treatment carries a high risk of entering small, critical arteries

*Cerebral Dural Arteriovenous Fistulas*
https://doi.org/10.1016/B978-0-12-819525-3.00010-1

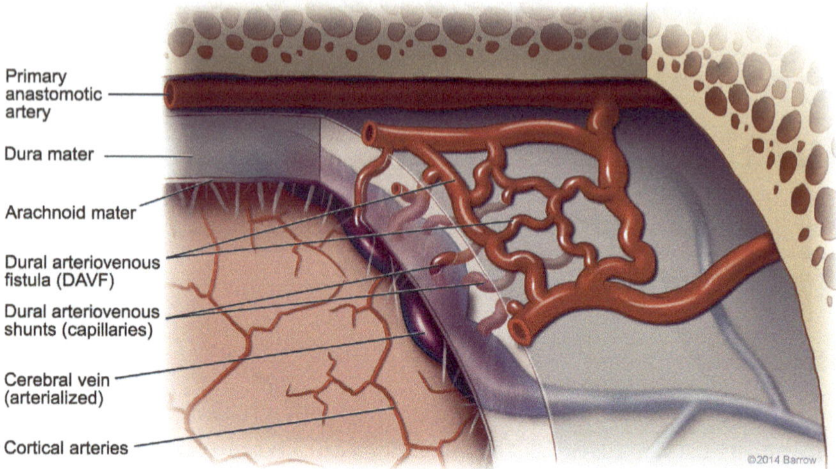


**Fig. 1** **Illustration of dural arteriovenous fistula angioarchitecture. The drawing illustrates the existing capillary networks that are recruited in pathological states.** *(Used with permission from Barrow Neurological Institute, Phoenix, Arizona.)*

(e.g., embolization of the central retinal artery while treating an ethmoidal dAVF), open surgical treatment should be considered. Detailed evaluation of venous-phase angiography is necessary to ensure that the planned target of venous outflow is not physiologic and is only a draining pathologic nidus. If not used cautiously, this procedure can lead to venous hypertension and venous infarction.

Galenic, straight sinus, and tentorial dAVFs are problematic to cure using an endovascular treatment.[7] These dAVFs have extensive arterial supplies from the meningeal arteries, which branch from the ICAs and vertebral arteries. They are hard to cannulate and pose a higher risk to embolize than ECA feeders. In addition, navigation to deep venous locations around the tentorium is difficult. Furthermore, these dAVFs often empty solely to subarachnoid veins instead of their associated sinus (Type III), which prevents transvenous access.[7] Therefore, the management of these dAVFs likely requires microsurgical treatment in contrast to most other dAVFs.[7–15]

## Open surgical considerations

We recommend that our patients with dAVFs and cortical venous reflux (CVR) undergo surgery when endovascular therapy fails, is not technically feasible, or is unfavorable due to anatomic constraints. Surgical goals can

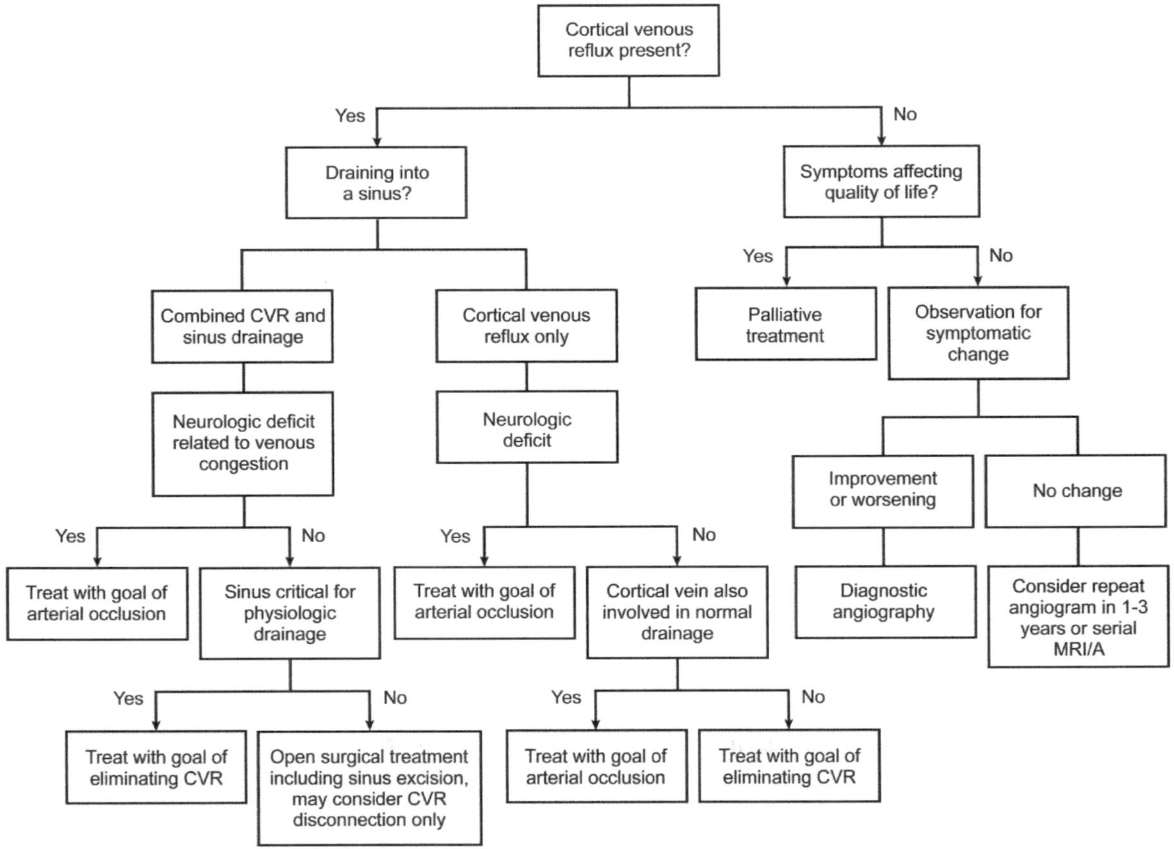


**Fig. 2** Flowchart of suggested management strategies based on anatomic and clinical dural arteriovenous fistula characteristics. *CVR*, cortical venous reflux; *MRI/A*, magnetic resonance imaging angiography. *(Used with permission from Barrow Neurological Institute, Phoenix, Arizona.)*

include obtaining venous access for embolization or direct packing of the fistula, surgical excision of the dAVF, or focal ligation of arterialized leptomeningeal veins (CVR) without removal of the nidus. For lesions with CVR only without sinus drainage, CVR disconnection alone is all that is required. Clinical application of this technique has been demonstrated to be effective while resulting in less surgical morbidity, and some groups prefer it to complete excision in many patients.[16–19] Before CVR disconnection, the neurosurgeon must decide whether it is safe to disrupt the veins exhibiting CVR. This decision is critical when the vein to be divided is a key cortical draining vein, such as the vein of Labbé. The venous phase of the cerebral angiograms must be studied in detail to ensure that the normal brain is not draining into the involved pathologic veins. In all cases, disruption of any arterialized draining veins should be performed as close as possible to the fistula nidus.

In the presence of thrombosis or stenosis of the dAVF venous outflow, cannulation to the fistula can be achieved by direct surgical exposure. The sinus can then be directly packed using coils or other thrombogenic materials such as gelatin foam or cellulose. This approach is also helpful for direct endovascular cannulation. If an orbital surgical exposure is needed, the surgeon may want to consider performing the procedure with an ophthalmologist present in case an emergent canthotomy is required for the development of a retro-orbital hematoma.

After the surgical interruption of arterial feeders and disconnection of arterialized leptomeningeal veins, the dura around the nidus is extensively coagulated and removed. Hemostatic clips can also be used for surgical obliteration. When accessing a deep lesion where minimal brain retraction is preferred, mannitol may be administered, and cerebrospinal fluid may be drained if feasible. Stereotactic navigation should be used for incision planning and intraoperative orientation when T2-weighted imaging demonstrates flow voids that can be used to guide access to the surgical corridor.

The extensive open surgical approaches used in the excision of complex dAVFs can lead to significant complications. The establishment of surgical corridors can be challenging when the lesion is located deep within the brain-like tentorial and posterior fossa lesions. Preparation for significant bleeding at any point in the procedure is important. Significant bleeding may occur that is related to distended vessels traversing the scalp and calvarium from external carotid sources, even on the initial exposure.[14]

## Dural arteriovenous fistula subtypes and treatment approaches

dAVFs can be categorized as supratentorial, tentorial, and posterior fossa. Some dAVF locations are associated with a more aggressive clinical course, such as those in the anterior cranial fossa or tentorium, as they have higher rates of CVR on presentation. This presentation relates to the structure of the venous anatomy in the region. For example, dAVFs in the anterior cranial fossa are more distal to large venous sinuses and therefore tend to drain into cortical veins.[20] We review the subtypes of dAVFs in these anatomic locations.

## Supratentorial dAVFs

### Anterior fossa/ethmoidal dAVFs

Any dAVF in the anterior cranial fossa region should be considered at high risk for hemorrhage. Because there are no large draining sinuses in this region, the venous outflow of anterior cranial fossa dAVFs is almost universally through cortical venous drainage into frontal veins. Transarterial embolization via the ophthalmic artery poses a risk of vision loss, and transvenous approaches are often not feasible. Given the favorable safety profile of surgical intervention and the difficulty of endovascular treatment, these dAVFs are often treated via open surgical means[21, 22] (Fig. 3). Surgical excision can be performed through either a traditional bifrontal craniotomy, or, often, via a unilateral frontal craniotomy. The former is preferred in the context of significant bilateral venous drainage.

### Superior sagittal sinus and convexity dAVFs

Superior sagittal sinus and other convexity dAVFs are well suited for surgical management (Fig. 4). A careful study of the preoperative angiography and noninvasive imaging studies, particularly MRI, is critical because eloquent cortical venous drainage needs to be assessed before treatment. Focal seizures or focal neurological deficits can trigger venous dilation, tortuosity, and mass effect.

Superior sagittal sinus dAVFs are usually supplied from middle meningeal artery branches. They can be approached via a craniotomy on the same side as the fistula. Rarely, bilateral craniotomies with two dural flaps anchored on the superior sagittal sinus are necessary to identify all feeding artery sources. In some patients without CVR, a high-flow dAVF may

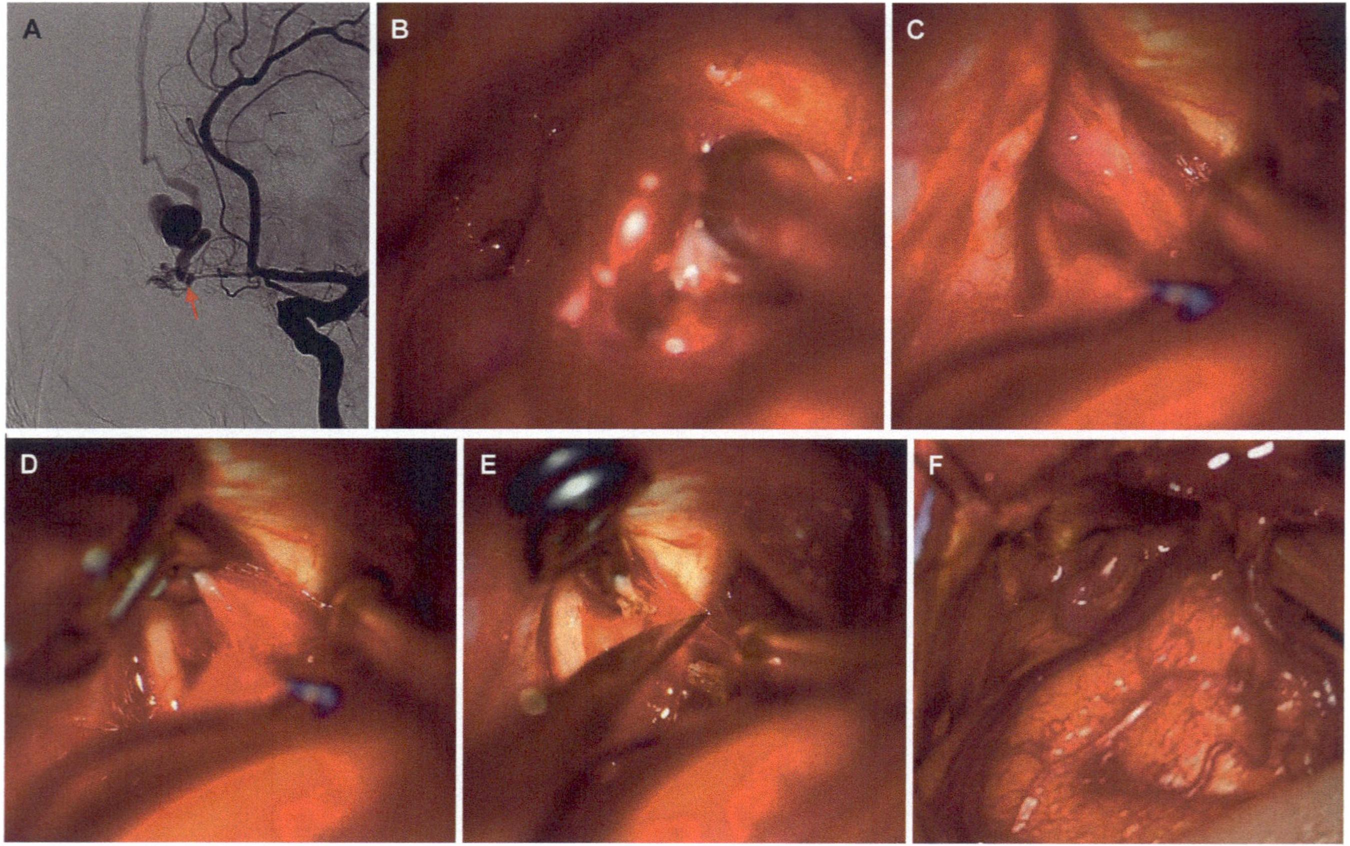


**Fig. 3** A 78-year-old man with incidental ethmoidal dural arteriovenous fistula (dAVF) found on evaluation for vertigo. (A) Oblique left internal carotid artery angiogram demonstrating feeding from an ophthalmic artery branch. The *red arrow* demonstrates the fistulous point with a cortical venous varix (Borden Type III dAVF). A bifrontal craniotomy was performed to access the fistula. (B) Initial surgical exposure with large cortical venous varix in view with an arterialized *reddish hue*. (C) Dissection and identification of fistulous point. (D) Clipping of fistulous point close to the dura. (E) Coagulation and ligation of draining vein. (F) View of draining vein after ligation, demonstrating a dearterialized *bluish hue*. *(Used with permission from Barrow Neurological Institute, Phoenix, Arizona.)*

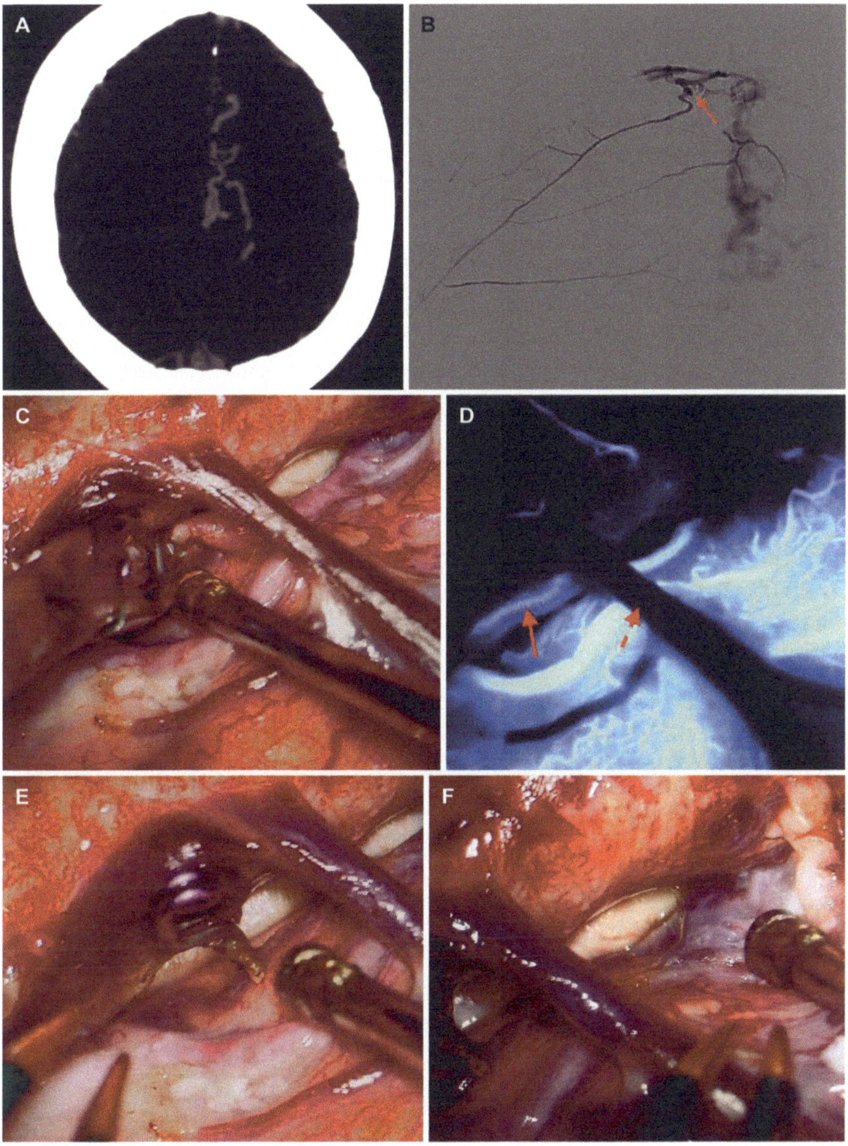


**Fig. 4** A 68-year-old man with seizures and unclear history of previous surgical resection of a vascular malformation. (A) Computed tomography angiogram demonstrating multiple venous varices near the left superior sagittal sinus. (B) Right middle meningeal artery selective angiogram with fistulous point (*red arrow*) near the superior sagittal sinus and drainage into the parafalcine cortical veins and the superior sagittal sinus (Borden Type II dural arteriovenous fistula). Embolization was performed, but a cure was not achieved. The patient then proceeded to open surgical treatment. (C) After performing a shallow interhemispheric dissection, multiple vessels were visible near the suspected fistulous site. (D) *Indocyanine green* angiography was performed to define the fistulous vessel. The interrupted *red arrow* indicates a physiologic draining vein, which is not fluorescent. The *solid red arrow* indicates a cortical vein with an arterialized flow from the superior sagittal sinus (fistulous point). (E) Clipping of fistulous point close to the dura, followed by coagulation and ligation of the draining vein. (F) View of draining vein after ligation, demonstrating a dearterialized *bluish hue. (Used with permission from Barrow Neurological Institute, Phoenix, Arizona.)*

shunt blood directly into a patent superior sagittal sinus. Such lesions are sometimes associated with intracranial hypertension or papilledema. They can be treated with substantial transarterial embolization or surgical skeletonization of the superior sagittal sinus to decrease the blood flow through the dAVF. Another common convexity variant is the sphenoparietal sinus dAVF. These dAVFs are typically classified as Type III fistulas. They can be treated through a frontotemporal craniotomy, which allows visualization of the sylvian draining veins that are arterialized with this fistula subtype. These veins can be followed to the dura and occluded with the use of a clip.

## Carotid cavernous dAVFs

Cavernous dAVFs account for approximately one-third of all dAVFs and can have ICA, ECA, or combined arterial blood supply.[19, 23] Barrow et al.[24] classified carotid cavernous fistulas into four groups. Type A fistulas are direct fistulas connecting the ICA and the cavernous sinus. Type A fistulas are not dAVFs. They are high-flow and are usually due to a laceration of the cavernous ICA or ruptured cavernous ICA aneurysm. These high-flow fistulas seldom resolve naturally. They require treatment to prevent the advancement of visual loss, insufferable bruit, or pain.[20] Most spontaneous carotid cavernous dAVFs are low-flow Types B, C, and D. They are fed by meningeal branches of the cavernous sinus. Type B fistulas originate from the ICA, Type C from the ECA, and Type D (most common) from both the ECA and the ICA. They typically do not have an identifiable cause and occur most commonly in middle-aged women. Patients typically present with ophthalmic symptoms, which can be bilateral.[25] In contrast to Type A fistulas, some of these low-flow fistulas undergo natural resolution.

CVR occurs in approximately 30% of patients with cavernous sinus dAVFs and is more common in patients with bilateral orbital signs.[19, 25–27] Cavernous sinus dAVFs without CVR may have a benign clinical course and can resolve over weeks or months. Although carotid cavernous fistulas without CVR are typically benign, some patients develop spontaneous venous thrombosis and venous hypertension, which can lead to increased intraocular pressure and increasing visual loss.[6] If medical therapy does not control ophthalmologic symptoms, endovascular treatment is indicated.[6]

If CVR is present, these dAVFs should be managed aggressively, typically with endovascular management via the inferior petrosal sinus, if patent or navigable, as the first approach.[28] Surgery can serve an adjunctive role, facilitating cannulation via the superior ophthalmic vein or the facial/angular veins. It can also be used for direct cannulation of cortical draining veins or surgical packing of the carotid cavernous fistula.[25, 29, 30] Continued monitoring after treatment is critical as venous thrombosis extending to the orbital veins can provoke venous hypertension and paradoxical visual decline and necessitate management with steroids or antithrombotic medications.[6]

Primary open microsurgical access of cavernous sinus dAVFs for direct ligation of a fistulous point was performed in the past with greater frequency than is currently done given improved modern endovascular techniques and favorable complication profiles associated with an endovascular approach.[31–34] Even with improving microsurgical techniques, complication rates with an open surgical approach remain relatively high. Multiple intradural and extradural approaches have been described, all of which can be greatly assisted with mannitol and lumbar drainage. The superior intradural approach involves first removing the anterior clinoid process and entering the cavernous sinus through the superior wall (either medially or laterally). This access can then be used for catheter embolization. For fistulas with a more anterior component, three extradural approaches can be taken; however, anterolateral and inferior approaches are generally more direct. The anterolateral approach allows extradural access to the floor of the anterior fossa and middle fossa for anterior clinoidectomy. After exposing the superior orbital fissure, the medial venous space and orbital veins are observed and can be used for catheter embolization. Regardless of the approach taken, only surgeons with extensive skull base experience should consider this an option, and only with other options unfeasible.

## Tentorial dAVFs

Tentorial dAVFs are particularly aggressive.[2, 7, 13, 35–38] Tentorial dAVFs have six broad subtypes: (1) galenic, (2) straight sinus, (3) torcular, (4) tentorial sinus, (5) superior petrosal sinus, and (6) incisural. The treatment strategy for these lesions can be decided systematically; we offer an optimal surgical approach that enhances gravity retraction, thereby allowing efficient subarachnoid dissection and fistula identification, for each tentorial dAVF subtype.[39]

### *Galenic dAVFs*

Galenic dAVFs are located in the midline at the posterior aspect of tentorial incisura and are associated with the vein of Galen. They have supratentorial or infratentorial drainage, or a combination of the two. The region near the vein of Galen is a deep corridor and can be difficult to navigate surgically. If surgery is necessary, lateral patient positioning and a torcular craniotomy with an interhemispheric approach provide the wide exposure required to decipher the various feeding and draining veins.[40] Once the dura is exposed and entered, gravity retraction is allowed to widen the operative field to facilitate the interhemispheric identification and dissection of the tangled feeding arteries and draining veins. The straight sinus is then skeletonized by cutting the falx superior to it and the tentorium bilaterally to the level of the straight sinus. These cuts allow visualization of the tentorium and both sides of the falx. For fistulas that drain inferiorly into the cerebellar veins, the sitting position, combined with a supracerebellar approach, allows good control of these draining veins. The tentorium can then be incised to expose a more superior view of the pertinent anatomy. The steep incline of the tentorial dura as it travels medially can limit the view of critical structures, particularly if there is extensive venous varicosity. Skeletonization has the additional benefit of disrupting ECA, tentorial, middle meningeal, and falcine feeding vessels. Any occipital artery supply is eliminated on initial exposure. Any posterior meningeal arteries can be identified on preoperative imaging and should be identified and coagulated on dural exposure.

Once the fistulous outflow veins are identified, they are clipped or coagulated. Of the five confluent veins of the Galenic complex, the precentral cerebellar veins and basal vein of Rosenthal are targeted most frequently for occlusion. The internal cerebral veins are prominent posteriorly, but rarely need to be occluded. The precentral cerebellar vein is typically located just outside the surgical view inferior to the falcotentorial junction, which can be mobilized after skeletonization for a better view. Intraoperatively, the relationship between the dura containing the fistula and the venous drainage that needs to be interrupted is not always clear.

Preoperatively, it is critical to determine if the straight sinus is patent. If it is occluded, the galenic trunk can be clip-occluded to interrupt the fistulous outflow. The exposure required for this occlusion is not extensive. In the case of a patent straight sinus with anterograde flow, the vein of Galen should be preserved. The only interrupted vessels should be the tributary branches of the vein of Galen. Care should be taken not to interrupt the main trunk. If the galenic tributary veins are blue and not arterialized, they may be left to maintain antegrade drainage from the normal deep cerebral circulation.

### Straight sinus dAVFs

dAVFs in the straight sinus drain into veins on the cerebellar-tentorial surface. For access to the supracerebellar infratentorial corridor, the patient is best placed in a sitting position. The semiprone "park bench" position can also be used. Since straight sinus dAVFs are more superficial than galenic dAVFs and usually have a single draining vein, skeletonization is typically not necessary. For a supracerebellar infratentorial approach, placing the patient in a prone position is often preferred to allow the surgeon to sit while operating. However, with an arm-bracing system, the surgeon can sit as long as the patient is in a vertical position with their neck flexed. This position relaxes and stabilizes the surgeon's arms for the relatively short duration procedure.

For adequate visualization of the inferior tentorium and superior surface of the cerebellum, the bone should be removed over at least the midportion of the torcular and the inferior half of the transverse sinus. Complications such as dural tears and venous sinus injuries are risky with the patient in the sitting position. Extreme caution should be used when lowering the head and flooding the field when an air embolus is suspected.

The arterialized draining vein can be identified on the surface of the cerebellum by its redder than usual appearance and its path into the dural fistula. The draining vein can also be seen exiting the dura from the subarachnoid space. It can be recognized by its thick wall, characteristic white color, and emanating vasa vasorum. The draining vein should be clipped where it exits the tentorial dura, then coagulated and cut.

### Torcular dAVFs

dAVFs in the torcular tend to be superficial and have supratentorial venous drainage. Type III fistulas can be managed by selectively interrupting the cortical draining veins. In the prone position, a torcular craniotomy is straightforward and does not require much dissection to identify the draining veins. Type II torcular fistulas, on the other hand, drain directly into the major sinuses. These sinuses must be preserved despite the interruption of the arterialized draining vein. Because less complete venous disruption is likely in cases of Type II torcular dAVFs, more extensive skeletonization may be necessary to find and carefully ligate the arterial feeders. The arterial feeders can converge into the torcular from any direction, including the falx, tentorium, and any dural surface converging on the torcular. These surfaces all require exploration and potential disruption.

### Tentorial sinus dAVFs

Classifications of the tentorial sinus anatomy have been undertaken by multiple groups without a clear consensus.[41–43] This region represents an area of considerably variable venous anatomy. In many cases, these are lesions of exclusion given a lack of association with other dural sinuses. They can reside laterally near the vein of Labbé or medially near the incisura.

These dAVFs have supratentorial drainage in most cases. Thus a supratentorial-infraoccipital approach is ideal given the broad exposure of the superior tentorial surface that this offers and allows quick identification and clipping of the fistulous vein. A prone or semiprone "park bench" positioning is preferred. The semiprone position with the head rotated toward the floor can help minimize central venous pressure, thus decreasing the extent of occipital lobe extrusion on dural opening. AP angiography should help to determine the craniotomy location. If the fistulous point is more medial, then a craniotomy incorporating the torcular and a wide dural opening assists in occipital lobe mobilization. For lateral tentorial dAVFs, a limited craniotomy near the transverse-sigmoid sinus junction will suffice.

### *Superior petrosal sinus dAVFs*

Superior petrosal sinus dAVFs typically drain into the superior petrosal vein. More radical transpetrosal approaches for open microsurgical treatment are unnecessary because only the exposure of the draining vein is required to interrupt the fistula. Although some clinicians conceptually group superior petrosal sinus dAVFs together with tentorial sinus dAVFs, they should be conceptualized separately given their location at the petrous ridge and involvement of the superior petrosal sinus.[23, 44] The ECA, ICA, and vertebrobasilar system provide the arterial supply via their separate dural branches. These dAVFs are typically associated with the closure of the ipsilateral superior petrosal sinus and resultant cortical venous drainage; patients with these lesions most commonly present with hemorrhage.[23, 44]

An extended retrosigmoid approach is appropriate for these lesions, allowing a full appreciation of the venous anatomy.[45] Cerebellar edema, often present with these lesions after hemorrhage, is most quickly decompressed by the quick release of cerebrospinal fluid from the cisterns. After cerebrospinal fluid release, the involved petrosal vein is readily visible, as it turns superiorly. Interruption with a surgical clip should be close to the temporal bone dura. If ancillary draining veins are also observed, they should be coagulated and cut. Any large venous varix should be mobilized to inspect for these smaller draining veins. Fistulous drainage may remain into the superior petrosal sinus, though as long as drainage in the superior petrosal vein is ligated, a Type I drainage pattern is an acceptable outcome requiring close follow-up.

### *Incisural dAVFs*

Like tentorial sinus dAVFs, these lesions are not well characterized in the literature, and both tend to drain supratentorially. Indeed, an incisural venous system is not present in most patients.[46] These dAVFs can also drain into the deep venous system and can extend posteriorly to involve the falx. The feeding arteries can be diverse and arise from ICA, ECA, posterior cerebral artery, or vertebral artery branches that traverse the dura. These dAVFs are nearly universally drained by the basal vein of Rosenthal, smaller lateral mesencephalic veins, or the vein of Galen. Subarachnoid drainage can also occur in cortical veins in either the superior or the inferior direction in addition to drainage into spinal medullary veins. As would be expected with this high-risk pattern of drainage, these lesions often present with bleeding or a neurological deficit. Spinal medullary involvement can induce myelopathy, and patients can present with progressive gait and upper extremity dysfunction.[47]

This dAVF type can produce a misleading angiogram and requires attention to the pattern of venous drainage. An incisural dAVF can easily be confused with a superior petrosal sinus dAVF, the latter of which requires an infratentorial approach. Incisural dAVFs are better managed through a posterior parasagittal craniotomy and interhemispheric approach, with a trajectory posterior to the splenium of the corpus callosum.[36] However, if the lesion appears to be closer to the uncus and supraclinoid ICA, a more anterior pterional-transsylvian approach or a lateral subtemporal approach should be considered. Surgical treatment consists, at a minimum, of CVR disconnection with complete resection as the goal.

## Posterior fossa dAVFs

Posterior fossa dAVFs are usually located in the transverse sinus or sigmoid sinus, inferior petrosal sinus, and the marginal sinus (foramen magnum). The recommendation to treat these dAVFs aggressively stands if CVR is present.[19, 37, 44] Surgical disconnection of the dAVF can be achieved with tolerable risk to the patient when endovascular management does not eradicate CVR.[19, 37]

### Transverse/sigmoid sinus dAVFs

With Type I transverse-sigmoid sinus dAVFs, the symptoms are typically confined to pulsatile tinnitus. In high-flow dAVF types with CVR and the communication of the transverse sinus with overall cranial drainage, global neurological symptoms related to venous hypertension can increase intracranial pressure and lead to venous infarcts and seizures. Arterial feeders typically include the middle meningeal artery, transosseous occipital branches, and the ascending pharyngeal artery along with tentorial branches of the ICA, and, at times, PCA. Venous drainage can be ipsilateral or contralateral if the ipsilateral transverse/sigmoid sinus is occluded.

Endovascular treatment can be attempted initially, with surgery reserved if treatment is unable to eliminate CVR in cases of Type II or III dAVF. Preoperative transarterial embolization should be used whenever possible to help minimize intraoperative bleeding. For the disconnection of this type of dAVF, the patient should be placed in the semiprone "park bench" position. The patient's head should be positioned with the side the lesion is on facing upward. An extended retrosigmoid approach with wide exposure of the transverse and sigmoid sinuses is ideal for these lesions. The occipital and posterior auricular arteries, which are usually enlarged, should be securely ligated and divided. In some cases, this is sufficient to dearterialize the dAVF. Once the dura is incised and opened, the arterialized draining vein is located. The vein is then clipped, coagulated, and cut.

When the goal of treatment is to manage CVR, then the dura is opened above or below the dAVF depending on the CVR drainage pattern noted on preoperative angiography. If there are any arterialized leptomeningeal veins, they should be readily identifiable, and all should be coagulated and divided. When the aim of treatment is to excise the fistula entirely along with the sacrifice of the transverse/sigmoid sinus, the mastoid and posterolateral portion of the temporal bone are drilled to reveal the dura laterally and anteriorly to the sigmoid sinus. Thorough knowledge and understanding of the anatomy of this surrounding area or assistance from an otolaryngologist are essential to preserving functional hearing and the facial nerve. Bleeding from arterial feeding branches that arise from the temporal bone can be controlled with bone wax. The transverse sinus is then skeletonized along a segment of at least a few centimeters, with feeding vessels cut and coagulated as the skeletonization proceeds. The transverse sinus can then be ligated. The occipital lobe can be elevated, and the cerebellum can be retracted to expose the tentorial dura, which should be coagulated and cut to disrupt any tentorial feeding arteries or cortical draining veins. The skeletonized sinus is removed distal to the transverse sigmoid junction. If the vein of Labbé is present, the patency of the sigmoid sinus should be preserved to maintain venous drainage from the brain. Ligation of the vein of Labbé can be considered if it drains the fistula exclusively because collateral venous drainage will have developed for that cortical territory.

## Inferior petrosal sinus dAVFs

Rare and benign, inferior petrosal sinus dAVFs tend to lack cortical venous drainage and therefore have a low hemorrhage risk. They typically have arterial supply from muscular and dural branches of the vertebral arteries along with an ECA supply arising from the middle meningeal, ascending pharyngeal, and occipital branches. Inferior petrosal sinuses can receive retrograde drainage from these dAVFs and flow into the cavernous sinus, producing ophthalmologic symptoms. Additionally, pulsatile tinnitus may arise from venous drainage flow into the ipsilateral jugular bulb. Venous drainage can even flow in a retrograde manner into the contralateral transverse sinus.[36] Although inferior petrosal sinus dAVFs are a challenge to expose surgically, they typically have good venous access. Therefore they are ideal candidates for endovascular treatment.

## Marginal sinus dAVFs

The marginal sinus encircles the foramen magnum and the cervicomedullary junction. It drains into the jugular veins or sigmoid sinuses. The marginal sinus communicates with the clival plexus anteriorly and the occipital sinus posteriorly. Arterial feeders of marginal sinus dAVFs frequently arise from the ascending pharyngeal artery, occipital artery, or vertebral meningeal arteries.[48] Typically the venous drainage of these dAVFs is toward the sigmoid sinus or jugular bulb via several smaller sinuses.[48]

Because the marginal sinus communicates with the clival plexus, some patients may present with ophthalmologic symptoms. These symptoms develop due to dAVFs that communicate with the cavernous sinus via the inferior petrosal sinus.[48] As with incisural dAVFs, these lesions can drain into spinal medullary veins and lead to progressive myelopathic symptoms.[49] These lesions are good candidates for endovascular therapy, with good venous and arterial access in most cases. Given the potential venous plexus anatomic variability and difficulty with hemostasis in this region, surgical intervention should be considered as a second-line treatment. If needed, the surgical approach most suitable for these lesions is a far-lateral corridor. Dural exposure typically occludes smaller feeding artery tributaries. With a wide dural opening, including the contralateral side, the dura can be elevated to visualize both the extradural posterior meningeal feeder and the intradural draining vein. This approach affords access to the vertebral artery to visualize the feeders that branch off its subarachnoid segment. A clip is then placed on the arterialized vein, and it is coagulated and cut as it exits the dura on top of the occipital condyle (Fig. 5).

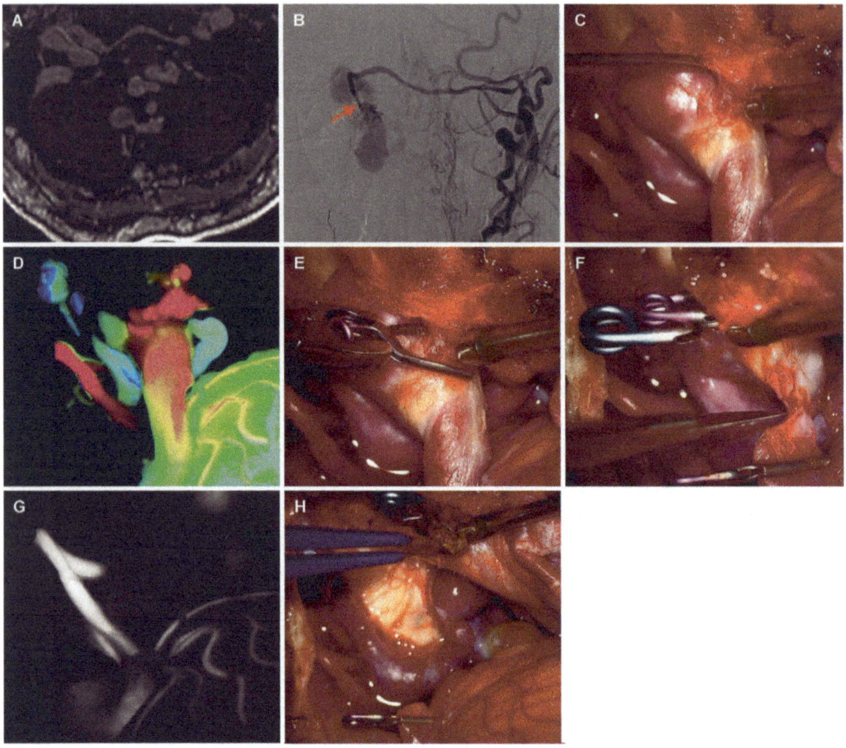


**Fig. 5** A 71-year-old man presented with left arm tingling and pain. (A) Axial T1-weighted magnetic resonance image with contrast, demonstrating multiple dilated venous varices in the inferior posterior fossa consistent with a marginal sinus dAVF with perimedullary drainage (Type III). (B) Left external carotid artery angiogram demonstrating a posterior meningeal feeding artery arising from the occipital artery, *red arrow* indicating fistulous point. A left far-lateral craniotomy was then performed to access the fistula. On dural opening, multiple dural feeding arteries were coagulated with bipolar cautery. (C) After opening the dura and dissecting the arachnoid, a midline fistulous point leading to a large draining vein complex was observed. (D) ICG angiography was performed to confirm fistulous flow. (E) Clipping of fistulous point close to the dura. (F) A branch of the draining vein was also clipped and cut to allow better mobilization of the fistulous point. (G) After improved dural mobilization was achieved, ICG angiography was performed to confirm the complete occlusion/ligation of the dAVF. (H) View of draining vein after ligation, demonstrating a dearterialized *bluish hue. dAVF,* dural arteriovenous fistula; *ICG*, indocyanine green. *(Used with permission from Barrow Neurological Institute, Phoenix, Arizona.)*

## Summary

Even with the rapid progress of endovascular treatment techniques, open surgical management remains critical in many dAVF subtypes as a primary treatment or supplement to endovascular management. The most important goal in any open surgical or endovascular treatment of dAVFs is the disruption of CVR, if present. Given the variety of cerebral dAVF locations and evolving treatment paradigms, a multidisciplinary approach incorporating surgery, transarterial embolization, transvenous embolization, or a combination is ideal. With any treatment modality, CVR disconnection effectively eradicates the risk of future adverse events, and, in some instances, may result in less morbidity than complete excision of the dAVF.

## Acknowledgments

The authors thank the staff of Neuroscience Publications at Barrow Neurological Institute for assistance with manuscript preparation.

## Disclosures

None.

## Financial support

None.

## References

1. Al-Shahi R, Bhattacharya JJ, Currie DG, Papanastassiou V, Ritchie V, Roberts RC, et al. Prospective, population-based detection of intracranial vascular malformations in adults: the Scottish Intracranial Vascular Malformation Study (SIVMS). *Stroke* 2003;**34**(5):1163–9. https://doi.org/10.1161/01.STR.0000069018.90456.C9.
2. Awad IA, Little JR, Akarawi WP, Ahl J. Intracranial dural arteriovenous malformations: factors predisposing to an aggressive neurological course. *J Neurosurg* 1990;**72**(6):839–50. https://doi.org/10.3171/jns.1990.72.6.0839.
3. Herman JM, Spetzler RF, Bederson JB, Kurbat JM, Zabramski JM. Genesis of a dural arteriovenous malformation in a rat model. *J Neurosurg* 1995;**83**(3):539–45. https://doi.org/10.3171/jns.1995.83.3.0539.
4. Sarma D, ter Brugge K. Management of intracranial dural arteriovenous shunts in adults. *Eur J Radiol* 2003;**46**(3):206–20.
5. Terada T, Higashida RT, Halbach VV, Dowd CF, Tsuura M, Komai N, et al. Development of acquired arteriovenous fistulas in rats due to venous hypertension. *J Neurosurg* 1994;**80**(5):884–9. https://doi.org/10.3171/jns.1994.80.5.0884.
6. van Dijk JM, Willinsky RA. Venous congestive encephalopathy related to cranial dural arteriovenous fistulas. *Neuroimaging Clin N Am* 2003;**13**(1):55–72.
7. Borden JA, Wu JK, Shucart WA. A proposed classification for spinal and cranial dural arteriovenous fistulous malformations and implications for treatment. *J Neurosurg* 1995;**82**(2):166–79. https://doi.org/10.3171/jns.1995.82.2.0166.
8. Collice M, D'Aliberti G, Arena O, Solaini C, Fontana RA, Talamonti G. Surgical treatment of intracranial dural arteriovenous fistulae: role of venous drainage. *Neurosurgery* 2000;**47**(1):56–66 [discussion 66–7].
9. Goto K, Sidipratomo P, Ogata N, Inoue T, Matsuno H. Combining endovascular and neurosurgical treatments of high-risk dural arteriovenous fistulas in the lateral sinus and the confluence of the sinuses. *J Neurosurg* 1999;**90**(2):289–99. https://doi.org/10.3171/jns.1999.90.2.0289.

10. Hoh BL, Choudhri TF, Connolly Jr ES, Solomon RA. Surgical management of high-grade intracranial dural arteriovenous fistulas: leptomeningeal venous disruption without nidus excision. *Neurosurgery* 1998;**42**(4):796–804 [discussion 804–5].
11. Kattner KA, Roth TC, Giannotta SL. Cranial base approaches for the surgical treatment of aggressive posterior fossa dural arteriovenous fistulae with leptomeningeal drainage: report of four technical cases. *Neurosurgery* 2002;**50**(5):1156–60 [discussion 1160–1].
12. Kiyosue H, Hori Y, Okahara M, Tanoue S, Sagara Y, Matsumoto S, et al. Treatment of intracranial dural arteriovenous fistulas: current strategies based on location and hemodynamics, and alternative techniques of transcatheter embolization. *Radiographics* 2004;**24**(6):1637–53. https://doi.org/10.1148/rg.246045026.
13. Lewis AI, Tomsick TA, Tew Jr JM. Management of tentorial dural arteriovenous malformations: transarterial embolization combined with stereotactic radiation or surgery. *J Neurosurg* 1994;**81**(6):851–9. https://doi.org/10.3171/jns.1994.81.6.0851.
14. Sundt Jr TM, Piepgras DG. The surgical approach to arteriovenous malformations of the lateral and sigmoid dural sinuses. *J Neurosurg* 1983;**59**(1):32–9. https://doi.org/10.3171/jns.1983.59.1.0032.
15. Ushikoshi S, Houkin K, Kuroda S, Asano T, Iwasaki Y, Miyasaka K, et al. Surgical treatment of intracranial dural arteriovenous fistulas. *Surg Neurol* 2002;**57**(4):253–61.
16. Al-Mahfoudh R, Kirollos R, Mitchell P, Lee M, Nahser H, Javadpour M. Surgical disconnection of the cortical venous reflux for high-grade intracranial dural arteriovenous fistulas. *World Neurosurg* 2015;**83**(4):652–6. https://doi.org/10.1016/j.wneu.2014.12.025.
17. Collice M, D'Aliberti G, Talamonti G, Branca V, Boccardi E, Scialfa G, et al. Surgical interruption of leptomeningeal drainage as treatment for intracranial dural arteriovenous fistulas without dural sinus drainage. *J Neurosurg* 1996;**84**(5):810–7. https://doi.org/10.3171/jns.1996.84.5.0810.
18. da Costa LB, Terbrugge K, Farb R, Wallace MC. Surgical disconnection of cortical venous reflux as a treatment for Borden type II dural arteriovenous fistulae. *Acta Neurochir* 2007;**149**(11):1103–8 [discussion 1108] https://doi.org/10.1007/s00701-007-1316-9.
19. van Dijk JM, TerBrugge KG, Willinsky RA, Wallace MC. Selective disconnection of cortical venous reflux as treatment for cranial dural arteriovenous fistulas. *J Neurosurg* 2004;**101**(1):31–5. https://doi.org/10.3171/jns.2004.101.1.0031.
20. Davies MA, TerBrugge K, Willinsky R, Coyne T, Saleh J, Wallace MC. The validity of classification for the clinical presentation of intracranial dural arteriovenous fistulas. *J Neurosurg* 1996;**85**(5):830–7. https://doi.org/10.3171/jns.1996.85.5.0830.
21. Gross BA, Moon K, Kalani MY, Albuquerque FC, McDougall CG, Nakaji P, et al. Clinical and anatomic insights from a series of ethmoidal dural arteriovenous fistulas at the Barrow Neurological Institute. *World Neurosurg* 2016;**93**:94–9.
22. Lawton MT, Chun J, Wilson CB, Halbach VV. Ethmoidal dural arteriovenous fistulae: an assessment of surgical and endovascular management. *Neurosurgery* 1999;**45**(4):805–10 [discussion 810–1].
23. Malek AM, Halbach VV, Higashida RT, Phatouros CC, Meyers PM, Dowd CF. Treatment of dural arteriovenous malformations and fistulas. *Neurosurg Clin N Am* 2000;**11**(1):147–66. ix.
24. Barrow DL, Spector RH, Braun IF, Landman JA, Tindall SC, Tindall GT. Classification and treatment of spontaneous carotid-cavernous sinus fistulas. *J Neurosurg* 1985;**62**(2):248–56. https://doi.org/10.3171/jns.1985.62.2.0248.
25. Meyers PM, Halbach VV, Dowd CF, Lempert TE, Malek AM, Phatouros CC, et al. Dural carotid cavernous fistula: definitive endovascular management and long-term follow-up. *Am J Ophthalmol* 2002;**134**(1):85–92.
26. Stiebel-Kalish H, Setton A, Berenstein A, Kalish Y, Nimii Y, Kupersmith MJ. Bilateral orbital signs predict cortical venous drainage in cavernous sinus dural AVMs. *Neurology* 2002;**58**(10):1521–4.

27. van Dijk JM, terBrugge KG, Willinsky RA, Wallace MC. Clinical course of cranial dural arteriovenous fistulas with long-term persistent cortical venous reflux. *Stroke* 2002;**33**(5):1233–6.
28. Benndorf G, Bender A, Lehmann R, Lanksch W. Transvenous occlusion of dural cavernous sinus fistulas through the thrombosed inferior petrosal sinus: report of four cases and review of the literature. *Surg Neurol* 2000;**54**(1):42–54.
29. Klisch J, Huppertz HJ, Spetzger U, Hetzel A, Seeger W, Schumacher M. Transvenous treatment of carotid cavernous and dural arteriovenous fistulae: results for 31 patients and review of the literature. *Neurosurgery* 2003;**53**(4):836–56 [discussion 856–7].
30. Miller NR, Monsein LH, Debrun GM, Tamargo RJ, Nauta HJ. Treatment of carotid-cavernous sinus fistulas using a superior ophthalmic vein approach. *J Neurosurg* 1995;**83**(5):838–42. https://doi.org/10.3171/jns.1995.83.5.0838.
31. Day JD, Fukushima T. Direct microsurgery of dural arteriovenous malformation type carotid-cavernous sinus fistulas: indications, technique, and results. *Neurosurgery* 1997;**41**(5):1119–24 [discussion 1124–6].
32. Krisht AF, Burson T. Combined pretemporal and endovascular approach to the cavernous sinus for the treatment of carotid-cavernous dural fistulae: technical case report. *Neurosurgery* 1999;**44**(2):415–8.
33. Mullan S. Treatment of carotid-cavernous fistulas by cavernous sinus occlusion. *J Neurosurg* 1979;**50**(2):131–44. https://doi.org/10.3171/jns.1979.50.2.0131.
34. Tu YK, Liu HM, Hu SC. Direct surgery of carotid cavernous fistulae and dural arteriovenous malformations of the cavernous sinus. *Neurosurgery* 1997;**41**(4):798–805 [discussion 805–6].
35. Cognard C, Gobin YP, Pierot L, Bailly AL, Houdart E, Casasco A, et al. Cerebral dural arteriovenous fistulas: clinical and angiographic correlation with a revised classification of venous drainage. *Radiology* 1995;**194**(3):671–80. https://doi.org/10.1148/radiology.194.3.7862961.
36. Lewis AI, Rosenblatt SS, Tew Jr JM. Surgical management of deep-seated dural arteriovenous malformations. *J Neurosurg* 1997;**87**(2):198–206. https://doi.org/10.3171/jns.1997.87.2.0198.
37. Tomak PR, Cloft HJ, Kaga A, Cawley CM, Dion J, Barrow DL. Evolution of the management of tentorial dural arteriovenous malformations. *Neurosurgery* 2003;**52**(4):750–60 [discussion 760–2].
38. Zink WE, Meyers PM, Connolly ES, Lavine SD. Combined surgical and endovascular management of a complex posttraumatic dural arteriovenous fistula of the tentorium and straight sinus. *J Neuroimaging* 2004;**14**(3):273–6. https://doi.org/10.1177/105122840426537.
39. Lawton MT, Sanchez-Mejia RO, Pham D, Tan J, Halbach VV. Tentorial dural arteriovenous fistulae: operative strategies and microsurgical results for six types. *Neurosurgery* 2008;**62**(3 Suppl. 1):110–24 [discussion 124–5] https://doi.org/10.1227/01.neu.0000317381.68561.b0.
40. Chi JH, Lawton MT. Posterior interhemispheric approach: surgical technique, application to vascular lesions, and benefits of gravity retraction. *Neurosurgery* 2006;**59**(1 Suppl. 1), ONS41-9 [discussion ONS41–9] https://doi.org/10.1227/01.NEU.0000219880.66309.85.
41. Matsushima T, Suzuki SO, Fukui M, Rhoton Jr AL, de Oliveira E, Ono M. Microsurgical anatomy of the tentorial sinuses. *J Neurosurg* 1989;**71**(6):923–8. https://doi.org/10.3171/jns.1989.71.6.0923.
42. Miabi Z, Midia R, Rohrer SE, Hoeffner EG, Vandorpe R, Berk CM, et al. Delineation of lateral tentorial sinus with contrast-enhanced MR imaging and its surgical implications. *AJNR Am J Neuroradiol* 2004;**25**(7):1181–8.
43. Muthukumar N, Palaniappan P. Tentorial venous sinuses: an anatomic study. *Neurosurgery* 1998;**42**(2):363–71. https://doi.org/10.1097/00006123-199802000-00097.

44. Ng PP, Halbach VV, Quinn R, Balousek P, Caragine LP, Dowd CF, et al. Endovascular treatment for dural arteriovenous fistulae of the superior petrosal sinus. *Neurosurgery* 2003;**53**(1):25–32 [discussion 32–3].
45. Quinones-Hinojosa A, Chang EF, Lawton MT. The extended retrosigmoid approach: an alternative to radical cranial base approaches for posterior fossa lesions. *Neurosurgery* 2006;**58**(4 Suppl. 2), ONS-208–14 [discussion ONS-214] https://doi.org/10.1227/01.NEU.0000192714.15356.08.
46. Picard L, Bracard S, Islak C, Roy D, Moreno A, Marchal JC, et al. Dural fistulae of the tentorium cerebelli. Radioanatomical, clinical and therapeutic considerations. *J Neuroradiol* 1990;**17**(3):161–81.
47. Ricolfi F, Manelfe C, Meder JF, Arrue P, Decq P, Brugieres P, et al. Intracranial dural arteriovenous fistulae with perimedullary venous drainage. Anatomical, clinical and therapeutic considerations. *Neuroradiology* 1999;**41**(11):803–12.
48. McDougall CG, Halbach VV, Dowd CF, Higashida RT, Larsen DW, Hieshima GB. Dural arteriovenous fistulas of the marginal sinus. *AJNR Am J Neuroradiol* 1997;**18**(8):1565–72.
49. Pierot L, Chiras J, Meder JF, Rose M, Rivierez M, Marsault C. Dural arteriovenous fistulas of the posterior fossa draining into subarachnoid veins. *AJNR Am J Neuroradiol* 1992;**13**(1):315–23.

# CHAPTER 9

# Hybrid surgical and endovascular treatment

**Joshua S. Catapano, Tyler S. Cole, and Felipe C. Albuquerque**
Department of Neurosurgery, Barrow Neurological Institute, St. Joseph's Hospital and Medical Center, Phoenix, AZ, United States

Dural arteriovenous fistulas (dAVFs) are pathologic, low-resistance cerebrovascular shunts between meningeal arteries and dural venous sinuses and/or cortical veins.[1] Because the annual risk of neurological deficit and/or hemorrhage associated with dAVFs is high, with approximately 4% risk of mortality associated with high-risk lesions with cortical venous reflux (CVR), treatment is often recommended if CVR is present or if symptoms affect the quality of life of the patient.[1,2] Both open surgical and endovascular approaches may be suitable depending on the location of the fistula. Endovascular therapy is the treatment of choice for the majority of lesions with drainage into a sinus with or without secondary reflux into a pial vein.[3] However, in challenging cases, a hybrid combined surgical and endovascular approach is warranted.[3–9] In particular, endovascular cases in patients with challenging vascular anatomy may benefit the most from a combined approach. For endovascular cases, depending on the location of the fistula, a transarterial or, more often, a transvenous approach is indicated; however, tortuous arterial and/or venous anatomy or venous stenosis/occlusion make such treatments via traditional access routes unfeasible.[3, 4, 6, 9–22] Herein, we summarize current hybrid approaches to address this issue. All patients gave informed consent for publication.

## Burr hole approach for draining sinuses

In cases in which a fistula is fed by limiting, small, tortuous arterial feeders, a transvenous approach is often necessary.[3, 23–25] Lesions with drainage directly into a sinus typically involve catheterization of the internal jugular vein to the draining sinus.[3] However, in instances with sinus thrombosis in which recanalization has not occurred, other treatment modalities are necessary.[3] Specifically, if the dAVF involves a superficial draining sinus, a

*Cerebral Dural Arteriovenous Fistulas*
https://doi.org/10.1016/B978-0-12-819525-3.00004-6

single burr hole or burr hole craniectomies may be used.[3] Houdart et al. has published the largest series to date with 10 patients (5 with transverse sinus dAVFs, 4 with superior sagittal sinus dAVFs, and 1 with a superior petrosal sinus dAVF) treated by a combined approach via burr hole craniectomies followed by catheterization of the sinus and successful embolization with coils and/or glue.[3] Other case reports and small case series have used similar techniques with successful embolization for these lesions.[10, 17, 21] We recommend the combined use of neuronavigation and angiography via transarterial access to locate the draining sinus for skin incision and creation of a burr hole craniectomy directly on the sinus, followed by puncturing of the sinus with a micropuncture needle under fluoroscopic guidance and confirmed with neuronavigation. Once access is achieved, a microwire is advanced under fluoroscopic guidance into the sinus, with removal of the needle followed by advancement of a micropuncture sheath. A microcatheter is then advanced over the wire, and embolization via coil, glue, and/or ethylene-vinyl alcohol copolymer (Onyx) can proceed.

## Burr hole or craniotomy for draining veins

In situations in which venous or arterial access is not possible and no superficial sinus is involved, a combined approach with either a burr hole or craniotomy for puncture of a draining vein may be feasible.[9, 11, 16, 20] Several case reports have demonstrated such procedures, with the majority describing such techniques for cavernous sinus fistulas without sufficient transvenous (including transophthalmic) access. In one iteration, a frontotemporal craniotomy was performed, and direct puncture of the superficial middle cerebral vein was obtained. The dAVF was then catheterized under fluoroscopic guidance and coil-embolized back into the distal superficial middle cerebral vein with obliteration of the fistula.[11, 16]

The combined approach of access via a draining vein can be useful beyond cavernous sinus lesions. Fig. 1A–E illustrates a case study performed at our institution involving a 57-year-old man who presented with intermittent expressive aphasia and blurry vision. Angiogram findings indicated that he had a Cognard Grade IV posterior parietal dAVF with arterial feeders from the external carotid artery via numerous branches of the left occipital artery and evidence of occlusion of the left transverse and sigmoid sinus. Retrograde cortical venous drainage was noted to involve the veins of Trolard and Labbé. The patient underwent angiography and partial occlusion of the fistula via Onyx embolization of one of the occipital

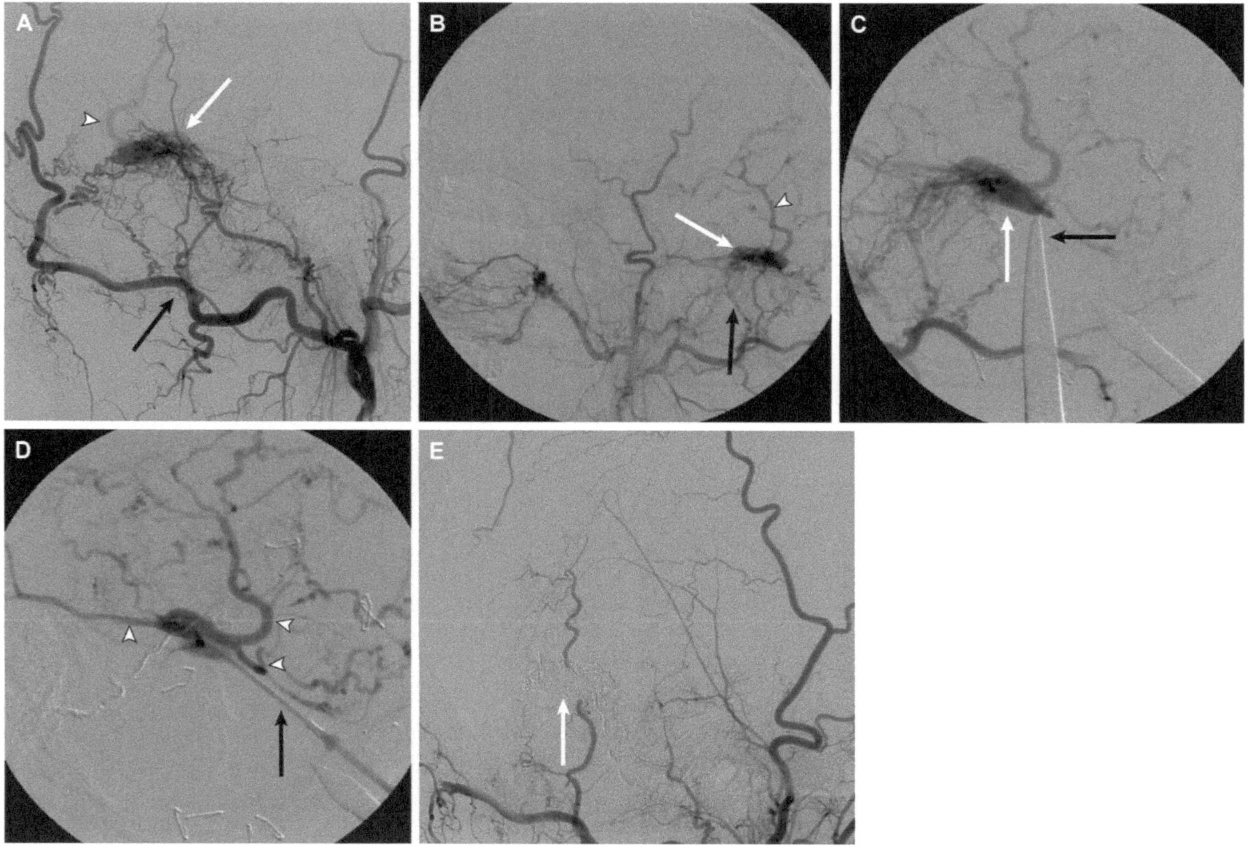


**Fig. 1** See figure legend on next page

artery branches. However, drainage into a large ectatic venous pouch was still apparent with cortical reflux into the veins of Trolard and Labbé. Direct puncture of the venous pouch for embolization was deemed necessary to fully treat this lesion. The patient underwent a subsequent angiogram via transfemoral arterial access and selective catherization of the left external carotid artery. After transfer to the operating room, the patient's head was placed in a Mayfield 3-point head fixation, and intraoperative neuronavigation was registered. Neuronavigation and angiography were used to locate the superficial venous pouch, and a linear incision was made over the pouch down to the scalp, where a burr hole craniectomy was performed. A 16-gauge angiocath needle was used to directly puncture the venous pouch, and a contrasted angiography run was performed to confirm presence in the pouch and reflux into the veins of Trolard and Labbé. A microcatheter was advanced into the pouch, and coil embolization was performed. Angiography findings revealed no further filling of the venous pouch and complete obliteration of the fistula. The patient recovered well from the procedure without neurological deficit.

## Direct venous puncture

Although performing direct venous puncture is technically difficult, it is effective and may be necessary for challenging, otherwise inaccessible fistulas. The majority of case reports and case series have described the successful direct puncturing of the superior ophthalmic vein for cavernous carotid fistulas when all other preferred transvenous and transarterial approaches were unsuccessful. Other studies have reported the direct puncturing of additional veins, including the facial vein, for successful embolization of these lesions.[26–40] Other authors have reported success in direct puncture of the

---

**Fig. 1** (A) Preembolization angiography of a Cognard Grade IV left posterior parietal dAVF with primary feeders from multiple branches of the left occipital artery (*black arrow*) and drainage into a venous ectatic pouch (*white arrow*) with cortical reflux (*white arrowhead*). (B) Following Onyx embolization of an occipital artery branch (*black arrow*), drainage into the venous pouch (*white arrow*) was still evident with cortical reflux (*white arrowhead*). (C) Angiography combined with stealth neuronavigation was used for precise localization of the venous pouch (*white arrow*) with a hemostat (*black arrow*). (D) After direct puncture of the venous pouch with a 16-gauge angiocath needle (*black arrow*), a pouchogram was performed that showed cortical venous reflux (*white arrowheads*). (E) Following coil embolization of the venous pouch (*white arrow*), complete occlusion of the fistula was found on angiogram. *(Used with permission from Barrow Neurological Institute, Phoenix, Arizona.)*

dAVF with combined approaches including the transorbital approach for cavernous sinus dAVFs and the foramen ovale approach for cavernous and clival dAVFs.[13, 14, 19, 22, 41]

Fig. 2 demonstrates an example of direct venous puncture. The patient was a 51-year-old woman who presented with mild proptosis and ecchymosis of her left eye of approximately 1-month duration. She experienced 2 days of acute

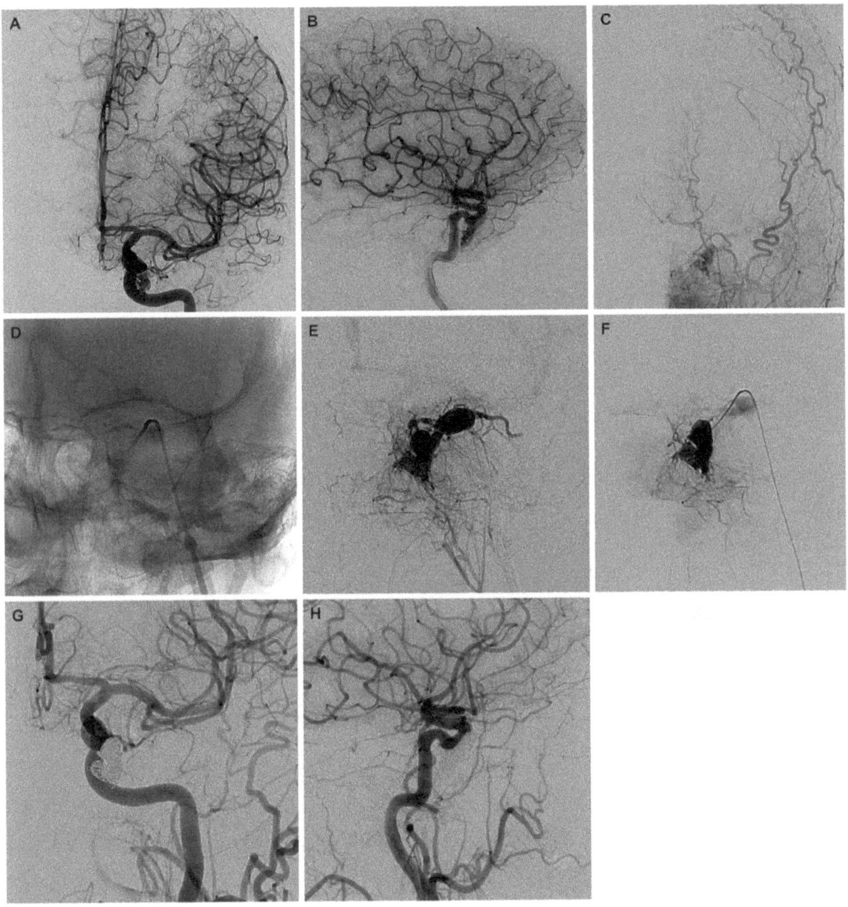


**Fig. 2** (A and B) Left internal carotid artery injection, posteroanterior and lateral view with multiple branches supplying dAVF from the left cavernous carotid artery and left external carotid artery injection (C) revealing distal branches of the internal maxillary artery supplying the dAVF. (D) Unsubtracted transorbital view with catheterization of left superior ophthalmic vein, followed by (E) a venous pouchogram and (F) a fistulogram showing the architecture of the dAVF. (G and H) Postcoiling left internal carotid artery injection posteroanterior and lateral view revealing complete dAVF obliteration. *(Used with permission from Barrow Neurological Institute, Phoenix, Arizona.)*

worsening of her left eye proptosis associated with vision changes, including decreased visual acuity and double vision on lateral gaze. She was evaluated at an outside hospital, where she was found to have a left indirect carotid cavernous fistula. Embolization was attempted but was unsuccessful because of lack of access to the fistulous pouch. Cerebral angiography with fistula embolization and possible direct superior ophthalmic vein access via a percutaneous puncture was offered to the patient. CT rotational angiography with a left external carotid artery contrast injection was performed with three-dimensional reconstructions revealing that the fistula could not be obliterated via a traditional transarterial or transvenous approach. There was not a safe target for embolization via a transarterial approach because the fistula had supply from multiple small branches of the left cavernous carotid artery and distal branches of the left internal maxillary artery. A traditional transvenous approach could not be performed because the portion of the cavernous sinus communicating with the inferior petrosal sinus did not communicate with the fistulous pouch. Furthermore, there was no access to the superior ophthalmic vein via the facial or angular veins secondary to a thrombus occluding the angular vein. Thus direct puncture of the superior ophthalmic vein was attempted. Initially the CT angiogram with a left external carotid artery injection was used to attempt to obtain a satisfactory trajectory for puncturing the superior ophthalmic vein. However, secondary to delayed contrast flow into the left superior ophthalmic vein, a satisfactory target was not visualized. Biplane angiography was then used to produce a venous roadmap, which guided the superior ophthalmic vein puncture. A small nick incision in the inferior eyebrow at the medial third of the superior orbit was made. With roadmap guidance, a micropuncture needle was used to gain access to the superior ophthalmic vein. Over a wire, a soft 4-French dilator sheath was gradually advanced. A superselective venogram confirmed placement of the sheath within the left superior ophthalmic vein. Through the sheath, a microcatheter over a microwire was navigated into the posterior aspect of the cavernous sinus. Superselective venography confirmed excellent microcatheter placement within the sinus. Multiple coils were then deployed, and subsequent angiography via the bilateral common carotid arteries demonstrated excellent obliteration of the fistula without evidence of complication.

## Craniotomy for arterial access

Although the majority of hybrid approaches to dAVFs are via venous routes, hybrid transarterial approaches are still effective and widely used for these lesions. For challenging cases in which a transarterial treatment is necessary but access is limited by tortuosity or other anatomical limitations, direct arterial access may be warranted. Several case reports in the literature have reported success with either craniotomies for access of feeding arteries (e.g., the middle meningeal artery) for embolization or direct external carotid artery catheterization for embolization of feeding arteries.[12, 15, 18] This is particularly relevant in cases of significant tortuosity or kinks in the proximal, dominant arterial feeder that precludes safe transarterial access in proximity to the fistula point.

## Conclusions

Although the majority of dAVFs can be treated via typical endovascular access routes or open microsurgical techniques, at times anatomic constraints and tortuous, limiting vascular anatomy necessitate the use of hybrid techniques. Surgical access to gain more direct access to the fistula should be considered in these cases. As reviewed in this chapter, surgical access can expose a targeted venous sinus, draining vein, or arterial feeder that can be directly cannulated for fistula embolization with standard neurointerventional techniques.

## Acknowledgments

The authors thank the Neuroscience Publications staff at Barrow Neurological Institute for helping with the manuscript preparation.

## Disclosures

None.

## Financial support

None.

## References

1. Reynolds MR, Lanzino G, Zipfel GJ. Intracranial dural arteriovenous fistulae. *Stroke* 2017;**48**(5):1424–31 [Epub 2017/04/23].
2. Zipfel GJ, Shah MN, Refai D, Dacey Jr RG, Derdeyn CP. Cranial dural arteriovenous fistulas: modification of angiographic classification scales based on new natural history data. *Neurosurg Focus* 2009;**26**(5):E14 [Epub 2009/05/05].
3. Houdart E, Saint-Maurice JP, Chapot R, et al. Transcranial approach for venous embolization of dural arteriovenous fistulas. *J Neurosurg* 2002;**97**(2):280–6 [Epub 2002/08/21].

4. Selvarajah E, Boet R, Laing A. Combined surgical and endovascular treatment of a posterior fossa dural arteriovenous fistula. *J Clin Neurosci* 2005;**12**(6):723–5 [Epub 2005/08/24].
5. Krisht AF, Burson T. Combined pretemporal and endovascular approach to the cavernous sinus for the treatment of carotid-cavernous dural fistulae: technical case report. *Neurosurgery* 1999;**44**(2):415–8 [Epub 1999/02/05].
6. Tsai LK, Liu HM, Jeng JS. Diagnosis and management of intracranial dural arteriovenous fistulas. *Expert Rev Neurother* 2016;**16**(3):307–18 [Epub 2016/02/03].
7. Nam TK, Byun JS, Choi HH, Chung MS, Lee EJ. Feasibility and effectiveness of direct puncture and onyx embolization for transverse sinus dural arteriovenous fistula. *Yonsei Med J* 2019;**60**(11):1112–5 [Epub 2019/10/23].
8. Park SH, Kim JH, Chang CH, Jung YJ. Fluoroscopy-guided combined (surgical/endovascular) treatment of dural arteriovenous fistula. *J Cerebrovasc Endovasc Neurosurg* 2017;**19**(2):106–10 [Epub 2017/11/21].
9. Shen SC, Tsuei YS, Chen WH, Shen CC. Hybrid surgery for dural arteriovenous fistula in the neurosurgical hybrid operating suite. *BMJ Case Rep* 2014;**2014** [Epub 2014/01/25].
10. Caplan JM, Kaminsky I, Gailloud P, Huang J. A single burr hole approach for direct transverse sinus cannulation for the treatment of a dural arteriovenous fistula. *J Neurointerv Surg* 2015;**7**(2):e5 [Epub 2014/01/17].
11. Chaudhary N, Lownie SP, Bussiere M, Pelz DM, Nicolle D. Transcortical venous approach for direct embolization of a cavernous sinus dural arteriovenous fistula: technical case report. *Neurosurgery* 2012;**70**(2 Suppl. Operative):343–8 [Epub 2011/11/11].
12. Crowley RW, Evans AJ, Jensen ME, Kassell NF, Dumont AS. Combined surgical/endovascular treatment of a complex dural arteriovenous fistula in 21-month-old. Technical note. *J Neurosurg Pediatr* 2009;**3**(6):501–6 [Epub 2009/06/03].
13. Dye JA, Buchanan CC, Gonzalez NR. Integrated open surgical and endovascular embolization treatment of a paracavernous venous plexus fistula: case report. *J Neurosurg* 2015;**122**(4):933–8 [Epub 2014/12/20].
14. Gil A, Lopez-Ibor L, Lopez-Flores G, Cuellar H, Murias E, Rodriguez-Boto G. Treatment of a carotid cavernous fistula via direct transovale cavernous sinus puncture. *J Neurosurg* 2013;**119**(1):247–51 [Epub 2013/05/15].
15. Hallaert GG, De Keukeleire KM, Vanhauwaert DJ, Defreyne L, Van Roost D. Intracranial dural arteriovenous fistula successfully treated by combined open-endovascular procedure. *J Neurol Neurosurg Psychiatry* 2010;**81**(6):685–9 [Epub 2010/02/24].
16. Hurley MC, Rahme RJ, Fishman AJ, Batjer HH, Bendok BR. Combined surgical and endovascular access of the superficial middle cerebral vein to occlude a high-grade cavernous dural arteriovenous fistula: case report. *Neurosurgery* 2011;**69**(2):E475–81 [discussion E81–2. Epub 2011/07/28].
17. Kong DS, Kwon KH, Kim JS, Hong SC, Jeon P. Combined surgical approach with intraoperative endovascular embolization for inaccessible dural arteriovenous fistulas. *Surg Neurol* 2007;**68**(1):72–7 [discussion 8. Epub 2007/06/26].
18. Lin N, Brouillard AM, Mokin M, et al. Direct access to the middle meningeal artery for embolization of complex dural arteriovenous fistula: a hybrid treatment approach. *J Neurointerv Surg* 2015;**7**(7):e24 [Epub 2014/06/12].
19. Liu A, Liu J, Qian Z, et al. Onyx embolization of cavernous sinus dural arteriovenous fistulas via direct transorbital puncture under the guidance of three-dimensional reconstructed skull image (reports of six cases). *Acta Neurochir* 2014;**156**(5):897–900 [Epub 2014/03/07].
20. Luo CB, Chang FC, Wu HM, Chung WY. Transcranial embolization of a transverse-sigmoid sinus dural arteriovenous fistula carried out through a decompressive craniectomy. *Acta Neurochir* 2007;**149**(2):197–200 [discussion. Epub 2006/11/09].

21. Pierot L, Visot A, Boulin A, Dupuy M. Combined neurosurgical and neuroradiological treatment of a complex superior sagittal sinus dural fistula: technical note. *Neurosurgery* 1998;**42**(1):194–7 [Epub 1998/01/27].
22. Puffer RC, Lanzino G, Cloft HJ. Using XperGuide planning software to safely guide catheter access to the cavernous sinus via transorbital puncture: a case report. *Neurosurgery* 2014;**10**(Suppl. 2):E370–3 [discussion E3. Epub 2014/02/19].
23. Halbach VV, Higashida RT, Hieshima GB, Goto K, Norman D, Newton TH. Dural fistulas involving the transverse and sigmoid sinuses: results of treatment in 28 patients. *Radiology* 1987;**163**(2):443–7 [Epub 1987/05/01].
24. Halbach VV, Higashida RT, Hieshima GB, Hardin CW, Pribram H. Transvenous embolization of dural fistulas involving the cavernous sinus. *AJNR Am J Neuroradiol* 1989;**10**(2):377–83 [Epub 1989/03/01].
25. Urtasun F, Biondi A, Casaco A, et al. Cerebral dural arteriovenous fistulas: percutaneous transvenous embolization. *Radiology* 1996;**199**(1):209–17 [Epub 1996/04/01].
26. Wolfe SQ, Cumberbatch NM, Aziz-Sultan MA, Tummala R, Morcos JJ. Operative approach via the superior ophthalmic vein for the endovascular treatment of carotid cavernous fistulas that fail traditional endovascular access. *Neurosurgery* 2010;**66**(6 Suppl. Operative):293–9 [discussion 9. Epub 2010/05/29].
27. Prochazka V, Cizek V, Kacirova R. Cavernous sinus dural fistula treated by transvenous facial vein approach. *Interv Neuroradiol* 2004;**10**(1):69–74 [Epub 2004/03/14].
28. Uflacker R, Lima S, Ribas GC, Piske RL. Carotid-cavernous fistulas: embolization through the superior ophthalmic vein approach. *Radiology* 1986;**159**(1):175–9 [Epub 1986/04/01].
29. Serratrice N, Baucher G, Reyre A, Brunel H, Fuentes S, Dufour H. Management of two cavernous sinus dural arteriovenous fistulae by direct microsurgical approach and catheterization of the superior ophthalmic vein. *Neurochirurgie* 2019;**65**(6):397–401 [Epub 2019/06/18].
30. Reis CV, Gonzalez FL, Zabramski JM, et al. Anatomy of the superior ophthalmic vein approach for direct endovascular access to vascular lesions of the orbit and cavernous sinus. *Neurosurgery* 2009;**64**(5 Suppl. 2):318–23 [discussion 23. Epub 2009/05/07].
31. Quinones D, Duckwiler G, Gobin PY, Goldberg RA, Vinuela F. Embolization of dural cavernous fistulas via superior ophthalmic vein approach. *AJNR Am J Neuroradiol* 1997;**18**(5):921–8 [Epub 1997/05/01].
32. Miller NR, Monsein LH, Debrun GM, Tamargo RJ, Nauta HJ. Treatment of carotid-cavernous sinus fistulas using a superior ophthalmic vein approach. *J Neurosurg* 1995;**83**(5):838–42 [Epub 1995/11/01].
33. Komiyama M, Morikawa K, Fu Y, Yagura H, Yasui T, Baba M. Indirect carotid-cavernous sinus fistula: transvenous embolization from the external jugular vein using a superior ophthalmic vein approach. A case report. *Surg Neurol* 1990;**33**(1):57–63 [Epub 1990/01/01].
34. Ha JG, Jeong HW, In HS, Choi SJ. Transvenous embolization of cavernous sinus dural arteriovenous fistula using the direct superior ophthalmic vein approach: a case report. *Neurointervention* 2011;**6**(2):100–3 [Epub 2011/11/30].
35. Gomez-Paz S, Vergara-Garcia D, Robinson M, Kicielinski KP, Thomas AJ, Ogilvy CS. Coil embolization of a carotid-cavernous fistula through superior ophthalmic venous access via external jugular vein puncture approach. *World Neurosurg* 2019;**131**:196 [Epub 2019/08/20].
36. Chalouhi N, Dumont AS, Tjoumakaris S, et al. The superior ophthalmic vein approach for the treatment of carotid-cavernous fistulas: a novel technique using Onyx. *Neurosurg Focus* 2012;**32**(5):E13 [Epub 2012/04/28].
37. Bulsara KR, Vasudevan R, Erdem E. Superior ophthalmic vein approach to carotid-cavernous fistulas. Technical note. *Neurosurg Focus* 2005;**18**(2):ECP3 [Epub 2005/02/18].

38. Briganti F, Caranci F, Leone G, et al. Endovascular occlusion of dural cavernous fistulas through a superior ophthalmic vein approach. *Neuroradiol J* 2013;**26**(5):565–72 [Epub 2013/11/10].
39. Berlis A, Klisch J, Spetzger U, Faist M, Schumacher M. Carotid cavernous fistula: embolization via a bilateral superior ophthalmic vein approach. *AJNR Am J Neuroradiol* 2002;**23**(10):1736–8 [Epub 2002/11/13].
40. Baldauf J, Spuler A, Hoch HH, Molsen HP, Kiwit JC, Synowitz M. Embolization of indirect carotid-cavernous sinus fistulas using the superior ophthalmic vein approach. *Acta Neurol Scand* 2004;**110**(3):200–4 [Epub 2004/08/03].
41. Urdaneta-Moncada A, Feng L, Chen J. Occlusion of a clival dural arteriovenous fistula using a novel approach through the foramen ovale. *J Neurointerv Surg* 2013;**5**(6):e46 [Epub 2012/11/29].

## CHAPTER 10

# Radiosurgery for cerebral dural arteriovenous fistulas

**Daniel A. Tonetti and L. Dade Lunsford**
Department of Neurological Surgery, University of Pittsburgh Medical Center, Pittsburgh, PA, United States

## Introduction

Intracranial dural arteriovenous fistulas (dAVFs) are abnormal arteriovenous shunts within the dura in which meningeal arteries shunt blood into dural sinuses or cortical veins. dAVFs cause symptoms dependent on their location and physiology and range from intracranial hemorrhage and nonhemorrhagic neurologic deficits to less-aggressive symptoms including tinnitus, headache, and ocular symptoms. Treatment options for dAVFs, as outlined in other chapters, include microsurgical disconnection, transarterial or transvenous endovascular embolization, and stereotactic radiosurgery (SRS). The choice of treatment option for a given patient with a newly diagnosed dAVF is dependent on a number of factors including clinical factors (patient history, time course of symptomatology, and physical examination findings), radiographic factors (presence of cortical venous reflux, anatomy of arterial feeders, and venous drainage), institutional factors (availability of surgical/endovascular/radiosurgical infrastructure, referral patterns, and surgeon preference), and symptomatology. Radiosurgery plays an important role in the management paradigm and treatment armamentarium of the modern cerebrovascular surgeon.

## Indications/patient selection

DAVFs have traditionally been stratified into "high risk" and "low risk" categories dependent on the presence or absence of cortical venous drainage (CVD); patients with CVD are more likely to present with nonhemorrhagic neurologic deficits (NHND) or frank intracranial hemorrhage.[1] Because of its immediacy and high cure rates, embolization is often the first-line treatment for high-risk dAVFs with surgery often serving as second-line treatment when the fistula site cannot be occluded or reached with embolisate[2–4]; SRS has been added in such patients if complete DAVF

*Cerebral Dural Arteriovenous Fistulas*
https://doi.org/10.1016/B978-0-12-819525-3.00002-2

closure was not achieved by embolization. Because of its low procedural morbidity SRS also has been used in conjunction with planned dAVF embolization as a means to achieve early closure (via embolization) and prevent delayed recanalization—a "belt and suspenders" approach.

However, accumulating natural history data suggests that outcome data for dAVFs with CVD should be stratified by the presenting symptom into "nonaggressive" or "aggressive" subgroups, with the latter referring to those patients who have sustained a prior NHND or hemorrhage.[5, 6] The rationale for treatment of aggressive dAVF is the prospective avoidance of hemorrhage or NHNDs, and the rationale for treatment of nonaggressive dAVF is the alleviation of symptoms (tinnitus, ocular symptoms, and headache). As such, SRS has traditionally largely been used for nonaggressive dAVFs. However, numerous natural history studies have demonstrated a significantly lower risk of hemorrhage from nonaggressive dAVF, despite the presence of CVD.[5, 7–9] As a result, SRS has also been used for patients with CVD but a nonaggressive presentation.[6]

Taken altogether, the indications for primary radiosurgical management of dAVF are debilitating symptoms or the presence of a residual shunt after other therapeutic techniques have been attempted. Such symptoms may include proptosis, chemosis, or diplopia in the context of carotid-cavernous fistulas, pulsatile tinnitus which may be relatable to a posterior fossa (e.g., transverse-sigmoid) fistula or headache. Radiosurgery for high-risk dAVF, including those which have previously hemorrhaged or present with an acute nonhemorrhagic neurologic deficit, should only be considered in patients for whom embolization or surgical treatment is considered of unacceptably dangerous risk; in modern clinical practice and given advances in endovascular techniques, this need has lessened.

SRS also has a role in the management of incompletely treated dAVFs after partial microsurgical excision or incomplete endovascular treatment; in these cases, SRS can be offered as an adjunctive or salvage therapy.[10] SRS has also been described as a complementary therapy in patients who undergo endovascular embolization.[11–16] It has been suggested that SRS can serve as a complementary treatment by increasing the likelihood of permanent dAVF obliteration. Pollock et al. found that symptoms resolved more quickly in patients who underwent both radiosurgery and embolization compared with patients who underwent only radiosurgery; as such, it has been suggested that an upfront combined endovascular/radiosurgical approach to decrease flow through the shunt may provide an immediate benefit to these patients while awaiting the beneficial effects of radiosurgery.[11]

At our institution, we perform embolization for patients with dAVF and features of increased hemorrhage risk (including most patients with CVD) and/or patients with rapid neurologic deterioration. When a dAVF can be completely embolized with angiographic obliteration and symptom resolution, radiosurgery should not be performed. The presence of CVD is not an absolute contraindication to SRS; those presenting with nonaggressive dAVF are candidates for SRS despite the presence or absence of CVD.[6] Based on our experience, we advocate radiosurgery alone as a safe alternative to surgery and/or embolization in clinically stable patients who can tolerate symptom resolution in a delayed fashion.

## Radiosurgical technique

The majority of clinical outcome studies and published techniques on SRS for dAVF have been utilizing the Gamma Knife (models U, B, C, 4C, or Perfexion), though reports of linear accelerator (LINAC)-based radiosurgery exist.[14, 17, 18] This chapter focuses on Gamma Knife radiosurgery.

Modern SRS treatment planning for cerebrovascular lesions is based on a combination of stereotactic magnetic resonance imaging (MRI) and high-resolution biplanar digital subtraction angiography (DSA). At our institution, patients first undergo stereotactic frame placement using mild conscious sedation followed by thin-sliced MRI after administration of gadolinium contrast. Sequences obtained include fast spin echo axial T2-weighted images in addition to three-dimensional spoiled gradient recalled acquisition in steady state (3D SPGR) images after contrast administration. Patients subsequently undergo biplane DSA of all suspected feeding arteries (including internal and external carotid injections) as the next step before radiosurgery.

Target localization is achieved through integration of imaging data from the thin-cut magnetic resonance imaging and catheter-based cerebral angiography. It is of critical importance to include all abnormal arteriovenous shunting within the leaflets of dura in targeting. Arterial feeders to the fistula proximal to the shunting site are not included in the target volume. Venous drainage that is distinct from the fistulous site is also not included. The volume of the target varies from a small focus point reflecting a simple arterial-venous shunt to the length of the venous sinus where multiple arterial inputs dump into the sinus.

Typical mean target margin dose is approximately 20 Gy (15–30 Gy).[19] The target coverage is typically prescribed at the 50% isodose line, since the fall off of dose is sharpest at this isodose using the Gamma Knife technology. Higher isodoses may be prescribed providing that the target volume is completely included in a dose that receives at least 20 Gy. Newer techniques in arteriovenous malformation SRS indicate that increasing the volume that receives a dose exceeding 22 Gy further increases obliteration rates. The number of isocenters used varies based on the type of fistula, its location, size, and the 3D conformality of dose delivered to the fistulous pouch.[20] The mean maximum dose is typically between 40 and 50 Gy.[19] Care must be taken to avoid excess radiation doses to nearby critical neural elements. In the case of carotid-cavernous fistulas, the maximum optic nerve point dose should not exceed 11 Gy and the average optic nerve dose (in patients without prior radiation) should be 8 Gy or less.

## Case examples

**(1)** A 60-year-old man presented with headache and was found on noninvasive imaging to harbor a Cognard Grade I tentorial dAVF supplied exclusively by branches of the left posterior meningeal artery. Ectatic veins from the fistula drained into the sinus at the torcula. Using Leksell GammaPlan software (Fig. 1) to integrate magnetic resonance imaging with angiography, the fistulous site was targeted with a marginal dose of 23 Gy and maximum dose 43 Gy (target volume 1.0 $cm^3$). Follow-up DSA at 36 months demonstrated complete obliteration (Fig. 2).

**(2)** A 75-year-old woman presented with pulsatile tinnitus and was subsequently referred for diagnostic angiography. DSA revealed a Cognard Type IIb dAVF with drainage into the left transverse sinus and cortical venous reflux, and she was referred for SRS. Imaging obtained at the time of radiosurgical treatment included selective external carotid angiography co-registered with magnetic resonance imaging (Fig. 3). The 1.9 $cm^3$ fistulous site was targeted with a marginal dose of 23 Gy and maximum dose 46 Gy.

## Follow-up strategy

The latency interval for dAVF obliteration after SRS is often 1–3 years.[20] With this in mind, we schedule the first clinical and radiographic (MRI) follow-up visit for 6 months after radiosurgery. After this time point, annual follow-up with complete neurologic clinical evaluation paired with magnetic resonance imaging are performed. If noninvasive imaging

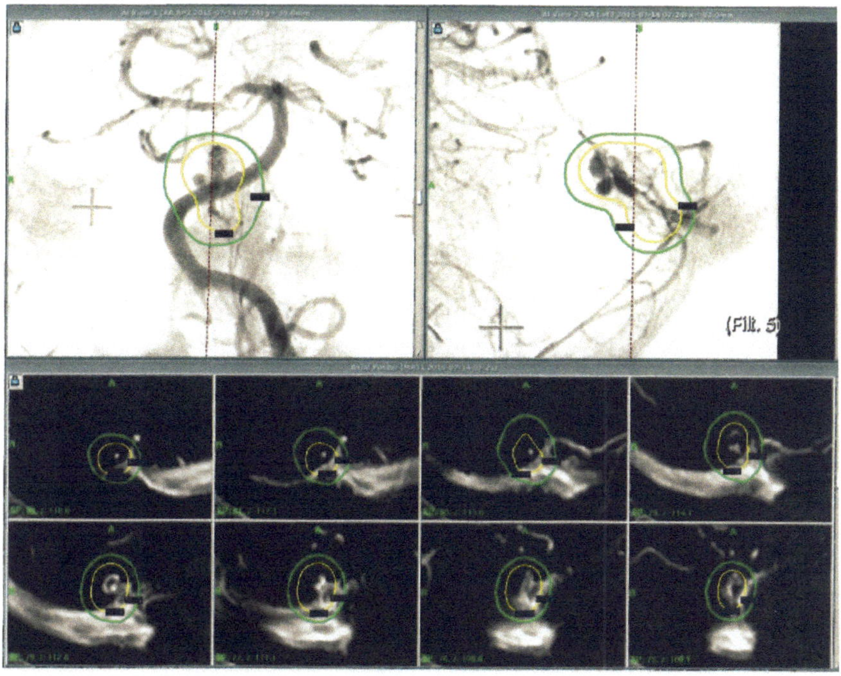

**Fig. 1** Diagram of Gamma Knife stereotactic radiosurgery planning in the Leksell GammaPlan software (AB Elekta, Stockholm, Sweden). The top panels display digital subtraction angiography of this dAVF. The bottom panel displays sequential thin-cut axial contrast-enhanced T1-weighted magnetic resonance imaging. The GammaPlan program, which contains co-registered cerebral angiography and MRI, illustrates a 50% isodose line (*yellow*) covering the target volume in all panels. In this example, the marginal dose was 23 Gy, maximum dose 43 Gy, and the target volume was 1.0 $cm^3$.

obtained 2–3 years after SRS is suggestive of complete dAVF obliteration, catheter-based angiography is performed to confirm obliteration. As a general rule, DSA is the preferred method as CTA has proven less satisfactory in defining obliteration outcome.

## Clinical and angiographic outcomes

Most studies detailing outcomes of SRS for dAVF have focused on angiographic obliteration[21,22] and/or symptomatic improvement as the outcomes of interest.[6,23] However, the rationale for treatment of high-risk versus low-risk dAVF is markedly different. For high-risk fistulas, the goal of treatment is the prospective avoidance of hemorrhage and/or NHND. For low-risk fistulas, the rationale for treatment is symptomatic improvement. As such, we believe that outcomes should be stratified by the primary goal of treatment.

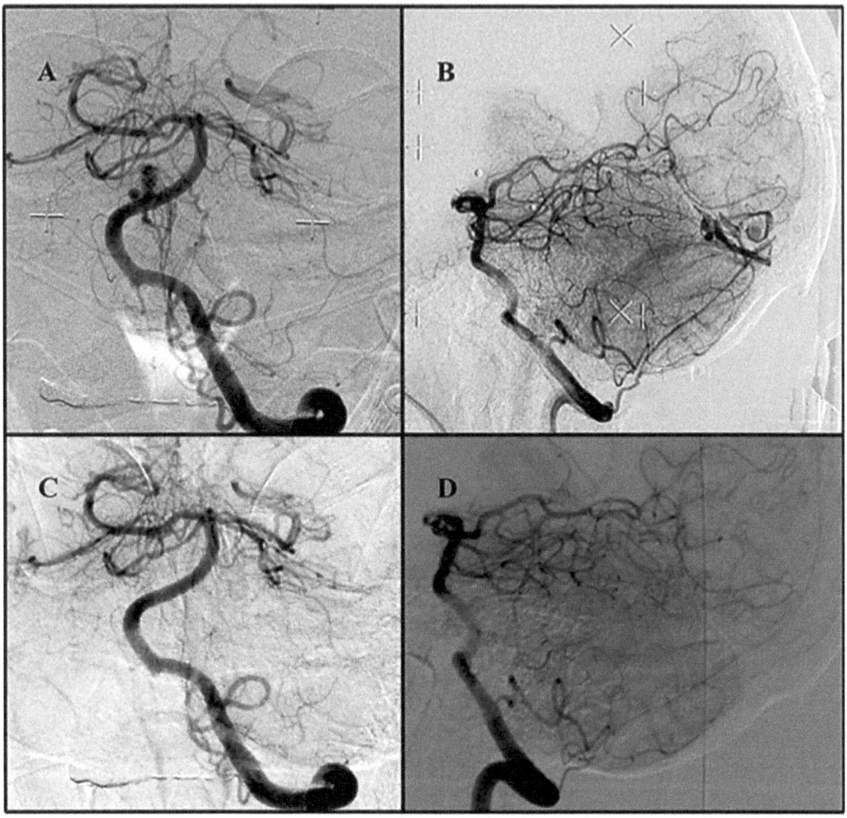


**Fig. 2** Anteroposterior (A) and lateral (B) views of digital subtraction angiography obtained at the time of radiosurgery demonstrate a Cognard Grade I dAVF fed by the left posterior meningeal artery with drainage into the torcula. Panels (C) and (D) are the same angiographic views obtained 36 months after SRS, demonstrating complete obliteration.

Regarding obliteration rates after radiosurgery, recent multicenter international consortium data have demonstrated actuarial obliteration rates at 3, 5, 7, and 10 years after radiosurgery as 41%, 61%, 70%, and 92%, respectively.[22] Predictors of obliteration in this study included the absence of venous ectasia, a carotid-cavernous location, and female sex.[22] Other predictors of obliteration after SRS have been reported including cavernous location,[24] low-risk dAVFs,[12, 25] small target volume.[25] Reports detailing predictors of obliteration are varied and sometimes contrasting; for example, cavernous location has been variably reported as associated with higher rates of obliteration[24] while other reports have not associated dAVF location with obliteration.[12,25] Similarly, gender was not associated with obliteration rates in several prior reports[12, 25] but was an independent predictor of obliteration in a multicenter consortium study.[22]

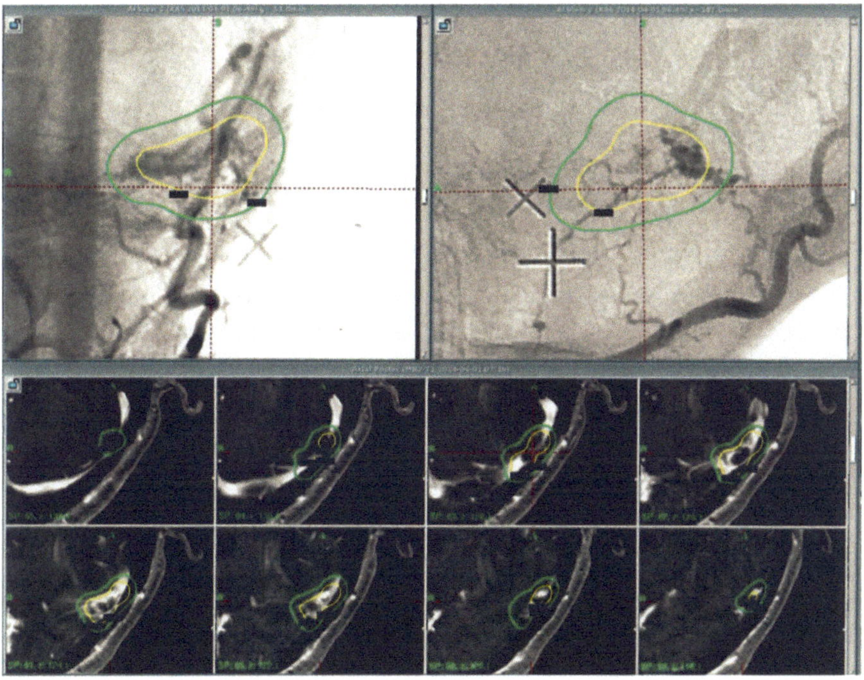

**Fig. 3** Diagram of Gamma Knife stereotactic radiosurgery planning in the Leksell GammaPlan software (AB Elekta, Stockholm, Sweden). A 50% isodose line (*yellow*) is shown covering the target volume in all panels. In this example, the marginal dose was 23 Gy, maximum dose 46 Gy, and the target volume was 1.9 $cm^3$.

In our experience, it is more likely that patient selection makes a larger difference on obliteration rate after SRS than does any individual characteristic, but high-flow dAVF (e.g., direct CCF or those dAVF with venous ectasia) are less likely to respond to SRS. This may be related to the likelihood of a larger nidus in these patients, which may increase the difficulty of accurate targeting; indeed, a larger nidus has been reported as being independently associated with SRS obliteration failure.[26] Venous ectasia may be an indicator of higher venous strain and thus may prevent radiation-induced luminal closure. Numerous studies have demonstrated higher rates of angiographic obliteration after SRS for dAVFs without CVD than those with CVD,[21, 25] which may also be related to increased flow through the fistulas with CVD.

dAVFs without cortical venous reflux do not pose a risk of hemorrhage or NHND and only very rarely go on to develop cortical venous reflux.[8, 27] As such, the meaningful rationale to treat these lesions for the patient and treating physician is symptomatic relief rather than angiographically confirmed obliteration. Pooled analyses of data available in the literature have demonstrated symptomatic improvement rates of approximately 98% for patients with tinnitus, and 90%–96% symptomatic improvement or resolution of ocular symptoms associated with indirect carotid-cavernous fistulas.[23]

## Hemorrhage rate after SRS

Historically, patients with aggressive and nonaggressive dAVF undergoing radiosurgical management have had their outcomes pooled. In these prior reports, rates of post-SRS hemorrhage during the latency interval for dAVF range from 0.4% to 5.0%.[12, 21, 28–30] However, dichotomizing outcomes based on presentation (and thus the goals of treatment) allows for a more appropriate risk/benefit analysis before treatment. Dichotomized by presentation, pooled postradiosurgery hemorrhage rates in data derived from international consortium studies, including those from our own institution, are approximately 0.8% per year—1.45% for Borden Grade 2–3 dAVF with aggressive presentation and exceedingly low (reported as 0%) for Borden Grade 2–3 dAVF with nonaggressive presentation.[31] Because the great majority of patients with aggressive presentation (hemorrhage or NHND) are treated endovascularly or surgically, these patients with aggressive fistulas treated with SRS likely reflect a biased subset of patients who either cannot tolerate a surgical intervention or harbor fistulas that were previously refractory to endovascular and/or surgical therapy, perhaps biasing these outcomes.

Data from the largest pooled outcome study demonstrated significantly lower rates of hemorrhage in the latency interval after SRS among patients with nonaggressive dAVFs when compared with aggressive dAVFs (annualized rate 0% vs 3.0% over 444.0 patient-years of follow-up, $P=.003$),[6] despite the presence of CVD. As such, radiosurgery is an important part of the treatment armamentarium of dAVF treatment for patients with nonaggressive presentation despite the presence of CVD.

## Conclusions

SRS is a useful tool in the multimodality treatment armamentarium for dAVFs, especially for nonaggressive dAVF, for patients who have failed before endovascular or surgical treatments, for patients who can tolerate delayed symptom resolution, or for those patients who are poor candidates for other treatment modalities.

## References

1. Awad IA, Little JR, Akarawi WP, Ahl J. Intracranial dural arteriovenous malformations: factors predisposing to an aggressive neurological course. *J Neurosurg* 1990;**72**(6):839–50.
2. Gross BA, Du R. Surgical treatment of high grade dural arteriovenous fistulae. *J Clin Neurosci* 2013;**20**(11):1527–32.
3. Liu JK, Dogan A, Ellegala DB, et al. The role of surgery for high-grade intracranial dural arteriovenous fistulas: importance of obliteration of venous outflow. *J Neurosurg* 2009;**110**(5):913–20.
4. Chandra RV, Leslie-Mazwi TM, Mehta BP, et al. Transarterial onyx embolization of cranial dural arteriovenous fistulas: long-term follow-up. *AJNR Am J Neuroradiol* 2014;**35**(9):1793–7.
5. Strom RG, Botros JA, Refai D, et al. Cranial dural arteriovenous fistulae: asymptomatic cortical venous drainage portends less aggressive clinical course. *Neurosurgery* 2009;**64**(2):241–7 [discussion 247–8].
6. Tonetti DA, Gross BA, Jankowitz BT, et al. Reconsidering an important subclass of high-risk dural arteriovenous fistulas for stereotactic radiosurgery. *J Neurosurg* 2018;**130**(3):972–6.
7. Soderman M, Pavic L, Edner G, Holmin S, Andersson T. Natural history of dural arteriovenous shunts. *Stroke* 2008;**39**(6):1735–9.
8. Gross BA, Du R. The natural history of cerebral dural arteriovenous fistulae. *Neurosurgery* 2012;**71**(3):594–602 [discussion 602–3].
9. Bulters DO, Mathad N, Culliford D, Millar J, Sparrow OC. The natural history of cranial dural arteriovenous fistulae with cortical venous reflux—the significance of venous ectasia. *Neurosurgery* 2012;**70**(2):312–8 [discussion 318–9].
10. Niranjan A, Lunsford LD. Stereotactic radiosurgery guidelines for the management of patients with intracranial dural arteriovenous fistulas. *Prog Neurol Surg* 2013;**27**:218–26.
11. Pollock BE, Nichols DA, Garrity JA, Gorman DA, Stafford SL. Stereotactic radiosurgery and particulate embolization for cavernous sinus dural arteriovenous fistulae. *Neurosurgery* 1999;**45**(3):459–66 [discussion 466–7].
12. Cifarelli CP, Kaptain G, Yen CP, Schlesinger D, Sheehan JP. Gamma knife radiosurgery for dural arteriovenous fistulas. *Neurosurgery* 2010;**67**(5):1230–5 [discussion 1235].
13. Friedman JA, Pollock BE, Nichols DA, Gorman DA, Foote RL, Stafford SL. Results of combined stereotactic radiosurgery and transarterial embolization for dural arteriovenous fistulas of the transverse and sigmoid sinuses. *J Neurosurg* 2001;**94**(6):886–91.
14. Lewis AI, Tomsick TA, Tew Jr JM. Management of tentorial dural arteriovenous malformations: transarterial embolization combined with stereotactic radiation or surgery. *J Neurosurg* 1994;**81**(6):851–9.
15. Link MJ, Coffey RJ, Nichols DA, Gorman DA. The role of radiosurgery and particulate embolization in the treatment of dural arteriovenous fistulas. *J Neurosurg* 1996;**84**(5):804–9.
16. Yang HC, Kano H, Kondziolka D, et al. Stereotactic radiosurgery with or without embolization for intracranial dural arteriovenous fistulas. *Neurosurgery* 2010;**67**(5):1276–83. discussion 1284–5.
17. Piippo A, Niemela M, van Popta J, et al. Characteristics and long-term outcome of 251 patients with dural arteriovenous fistulas in a defined population. *J Neurosurg* 2013;**118**(5):923–34.
18. Oh JT, Chung SY, Lanzino G, et al. Intracranial dural arteriovenous fistulas: clinical characteristics and management based on location and hemodynamics. *J Cerebrovasc Endovasc Neurosurg* 2012;**14**(3):192–202.

19. Koebbe CJ, Singhal D, Sheehan J, et al. Radiosurgery for dural arteriovenous fistulas. *Surg Neurol* 2005;**64**(5):392–8. discussion 398–9.
20. Yang HC, Lee CC, Pan DHC, Chung WY. Radiosurgery for dural arteriovenous fistulas. *Prog Neurol Surg* 2019;**34**:248–59.
21. Chen CJ, Lee CC, Ding D, et al. Stereotactic radiosurgery for intracranial dural arteriovenous fistulas: a systematic review. *J Neurosurg* 2015;**122**(2):353–62.
22. Starke RM, McCarthy DJ, Chen CJ, et al. Evaluation of stereotactic radiosurgery for cerebral dural arteriovenous fistulas in a multicenter international consortium. *J Neurosurg* 2019;1–8.
23. Tonetti DA, Gross BA, Jankowitz BT, et al. Stereotactic radiosurgery for dural arteriovenous fistulas without cortical venous reflux. *World Neurosurg* 2017;**107**:371–5.
24. Yang H, Kano H, Kondziolka D, et al. Stereotactic radiosurgery with or without embolization for intracranial dural arteriovenous fistulas. *Prog Neurol Surg* 2013;**27**:195–204.
25. Hanakita S, Koga T, Shin M, Shojima M, Igaki H, Saito N. Role of Gamma Knife surgery in the treatment of intracranial dural arteriovenous fistulas. *J Neurosurg* 2012;**117**(Suppl):158–63.
26. Yen CP, Ding D, Cheng CH, Starke RM, Shaffrey M, Sheehan J. Gamma Knife surgery for incidental cerebral arteriovenous malformations. *J Neurosurg* 2014;**121**(5):1015–21.
27. Satomi J, van Dijk JM, Terbrugge KG, Willinsky RA, Wallace MC. Benign cranial dural arteriovenous fistulas: outcome of conservative management based on the natural history of the lesion. *J Neurosurg* 2002;**97**(4):767–70.
28. Park KS, Kang DH, Park SH, Kim YS. The efficacy of gamma knife radiosurgery alone as a primary treatment for intracranial dural arteriovenous fistulas. *Acta Neurochir* 2016;**158**(4):821–8.
29. Soderman M, Edner G, Ericson K, et al. Gamma knife surgery for dural arteriovenous shunts: 25 years of experience. *J Neurosurg* 2006;**104**(6):867–75.
30. Wu HM, Pan DH, Chung WY, et al. Gamma Knife surgery for the management of intracranial dural arteriovenous fistulas. *J Neurosurg* 2006;**105**(Suppl):43–51.
31. Starke RM, McCarthy DJ, Chen CJ, et al. Hemorrhage risk of cerebral dural arteriovenous fistulas following Gamma Knife radiosurgery in a multicenter international consortium. *J Neurosurg* 2019;1–9.

# Author Index

Note: Page numbers followed by *t* indicate tables.

## K



## L

## Z

# Subject Index

Note: Page numbers followed by *f* indicate figures and *t* indicate tables.

FSC
www.fsc.org
MIX
Papier | Fördert gute Waldnutzung
FSC® C083411

Kolibri 360 GmbH
Ferdinand-Jühlke-Straße 7
99095 Erfurt, Deutschland
produktsicherheit@kolibri360.de